Reimagined

Building Products with Generative AI

First Edition

Shyvee Shi
Caitlin Cai
Dr. Yiwen Rong

peak
pioneer

Copyrighted Material

Reimagined: Building Products with Generative AI
Copyright ©2024 by Shyvee Shi, Caitlin Cai, Dr. Yiwen Rong, and PeakPioneer LLC.

All Rights Reserved.

No part of this publication may be reproduced, stored in a retrieval system or transmitted, in any form or by any means, electronic, mechanical, photocopy, recording or otherwise, without the written permissions from the authors, except for the inclusion of brief quotation in a review.

ISBN: 979-8-9899669-0-5

Notes to Readers

The views and opinions expressed in this book are solely ours and do not reflect those of our employers.

We have utilized GPT-4, Midjourney, and other AI tools for research, drafting, and refining. Recognizing the issues of credibility and trust with AI-generated content, we aimed to demonstrate the potential of building generative AI products through active collaboration with AI. This book is the result of a meticulous process, blending our original insights and extensive research with hundreds of AI prompting and iterations.

This book, addressing the rapidly evolving field of generative AI, is designed to stimulate current discussions and learning. The fast-paced nature of this technology means some information might soon be outdated. Our goal is to engage in a dynamic, ongoing conversation rather than presenting a definitive, static viewpoint. Your feedback is invaluable to us for continuous refinement and understanding.

Praise for *Reimagined*

"*Reimagined* is a valuable resource for product managers, helping them navigate the new AI landscape while emphasizing the fundamentals of successful product development." – **Tamar Yehoshua, IVP Venture Partner, Board member, ex-Chief Product Officer at Slack, VP at Google**

"*Reimagined* is a pivotal read for PMs, pushing the innovation envelope in generative AI. It expands the realms of imagination of what we can envision and achieve in product development." – **Amit Fulay, VP of Product Management, Microsoft Teams and GroupMe**

"*Reimagined* brilliantly surveys the power of AI in product development, offering in-depth strategies and practice examples that are essential for any builder creating truly intelligent products." – **Ravi Mehta, Co-founder and CEO at Outpace, Executive In Residence at Reforge, ex- Chief Product Officer at Tinder, and Product Leader at Facebook, TripAdvisor, and Xbox**

"*Reimagined* is the best guide to Generative AI I've found anywhere. Shi, Cai and Dr. Rong are at the forefront of this long-term trend. With Reimagined, you can be, too." – **Robert C. Wolcott, Adjunct Professor of Innovation at Booth and Kellogg, venture investor and Co-Founder of TWIN Global**

"*Reimagined* is no hype, it's a steady hand guiding you through the intricacies of AI. Shyvee, Caitlin, and Dr. Rong distill their diverse experience to illuminate possibilities with actionable strategies and real-world examples. An engaging read for AI explorers."- **Xing Yao, Founder of XVerse, AI + 3D Metaverse Unicorn Backed by Sequoia, Tencent, Temasek, and Hillhouse, Founder of Tencent AI Lab and Robotics X Lab**

"*Reimagined* is a MUST HAVE for product leaders navigating the dynamic era of AI. This guide is your key to understanding the transformative potential of AI, while also addressing real challenges and ethical considerations with the AI revolution." – **Jennifer Liu, Product Executive and Leadership Coach, ex-Senior Vice President of Product at Lattice, Chief Product Officer at Ethos Life, and Senior Director of Product Management at Google**

"The advent of Generative AI has significantly narrowed the gap between technological development and product innovation. *Reimagined* is a must have for entrepreneurs aspiring to thrive in the GenAI domain: the methodology to grasp the evolving key elements of product creation in the AI era, and the mindset to stay abreast of the constantly evolving AI technology." – **Dr. Yangqing Jia, Cofounder & CEO at Lepton AI. Creator of Caffe and Co-lead of PyTorch 1.0 and ONNX, ex-VP of Alibaba and President of Alibaba Cloud's Computing Platform, Director of Engineering at Facebook AI, Research Scientist at Google Brain**

"*Reimagined* cuts through the noise, offering product managers the crucial insights they need in this fast-evolving field." – **Laura Marino, Chief Product Officer, Advisor to CEOs and Heads of Product in high-growth companies TrueML, Lever, Board Member, Guest Lecturer at Stanford University**

"A must-have handbook for AI PMs and Entrepreneurs. It provides not only clear and actionable analytics frameworks, but also rich case studies about real-world business applications. Strongly recommended and awesome work, Shi, Cai and Dr. Rong!" – **Robert Dong, Generative AI Entrepreneur, Former Head of Product for Tiktok Creative AI, ex-AI product lead in Cruise automation and Meta**

"*Reimagined* is a must-read, masterfully blending AI insights with practical strategies for product development and ethical considerations. A game-changer for product managers and AI enthusiasts alike." – **Dr. Marily Nika, founder of the AI Product Academy and AI PM Bootcamp, AI Product Lead at Google/Meta**

"A must-read for product builders navigating the AI revolution, eager to harness the true power of generative AI to make a positive impact on the world." - **Tony Beltramelli, CEO & Co-founder Uizard, an AI-powered UI design tool**

"Generative AI will have a big impact on Product Management, both on the products itself and also on the way PMs are working. Based on my own work in the European Market I see a big demand for a book like this and I am curious to learn more from Shyvee and her coauthors." – **Raphael Leiteritz, Co-founder and Partner at Peak Product & Product Management Festival, former seasoned Google exec, advisor to Chief Product Officers, and angel investor to 200+ startups, including 20+ unicorns**

"A wonderful piece of work from Shi, Cai and Rong, summarizing the history, foundation, and go forward strategy for the next phases of generative AI. An important read for current and future product leaders looking to build in this space, and move this exciting industry forward." – **Erica Van, Partner at Intel Capital, Former Strategist at JPMorgan Chase & Co.**

"Shyvee and team bring an indispensable practitioner lens into navigating the generative AI product lifecycle - a first book of its kind to provide a strategic blueprint for our AI-powered future." – **Piyush Gupta, Chief Product Officer at Eva AI**

"*Reimagined* provides what's missing from most AI books: a pragmatic and practical approach to building great AI-enabled products and services" – **Phyl Terry, Founder at Collaborative Gain, Author of *Never Search Alone*, Co-Author of *Customers Included***

"*Reimagined* stands as an invaluable asset for product professionals navigating the fast-paced landscape of the Gen-AI era." – **Kai Yang, VP of Product Landing AI, AI entrepreneur & product executive**

"Caitlin is one of the earliest AI native Product and Go-To-Market leader that I have known before it is popular, and has always been in the frontier of creating AI native product with unique insights. *Reimagined* is a must read for product managers and founders working on Gen AI products and exploring GTM strategies." – **Bill Sun, Founder/CEO of GAlpha.ai; AI Researcher at Google Brain, Citadel and Cubist, Quant Portfolio Manager managed billion dollar GMV at Millennium Worldquant, Stanford Math PhD advised by David Donoho and Stephen Boyd (Head of BlackRock AI Labs)**

"Shyvee Shi doesn't just teach product management; she embodies it. Her wisdom is a testament to a career dedicated to mastering the art and science of building exceptional products." – **Lewis Lin, 8x Bestselling Author, CEO of ManageBetter and Impact Interview, former product leader at Google and Microsoft**

"Shyvee is a true lifelong learner, and in this book, she uses her deep curiosity and rigorous discipline to develop a powerful guidebook for product managers trying to get their arms around what generative AI means for them." – **Robbie Kellman Baxter, Founder of Peninsula Strategies, advisor to the world's subscription-based companies, keynote speaker, and author of The Membership Economy and The Forever Transaction**

"Shyvee is at the forefront of innovation within the PM community from her work interviewing top product leaders. This book emerges as essential reading for all product managers, offering a crucial guide on integrating AI effectively into their strategies and workflows." – **Pulkit Agrawal, Cofounder & CEO at Chameleon, the deepest product adoption platform for SaaS teams**

About the Authors

Shyvee Shi is a product management and innovation expert, currently serves as a Product Lead at LinkedIn and a top-rated instructor on LinkedIn Learning. She has impacted over 100K global followers and her content has reached 40M+ views. Her PM Learning Series has featured esteemed guests including the cofounders of LinkedIn and Google Docs/Sheets, former CPOs of Tinder, Slack, Yelp, Brex, Amplitude, CEO of Ancestry, and leading AI leaders from Roblox Studio, Uizard, Microsoft, Copy.ai, Synthesia, Forethought, and Google Language AI. As President of the Kellogg Alumni Club in San Francisco, Shyvee is a thought-after speaker and has influenced digital strategies at VMware, Disney, Cisco, Toyota, HSBC, Telstra, DFS, and many Fortune 1000 companies.

Caitlin Cai is a Wharton MBA and Northwestern alumnae, brought over a decade of experience in AI commercialization, go-to-market, product, and finance (private equity & venture capital), with first-hand experience in building AI companies and products with AI thought leaders Stanford AI Professor Dr. Andrew Ng (co-founder of Coursera and Google Brain), Xing Yao (the founder of Tencent AI Lab), and Robert Dong (former head of product Creative AI at Tiktok) in early founding teams. Ms. Cai has been an AI entrepreneur and has served on the advisory boards of multiple AI companies. Deeply passionate for Creative AI, Ms. Cai has been building AI products and companies to unleash human creativity and deepen connections.

Dr. Yiwen Rong is a seasoned product manager and a vanguard in the integration of AI into consumer and SaaS products. Currently working at Google as an AI product manager, he builds consumer products with AI that impacts billions of users' lives. Dr. Rong was a 2X entrepreneur and product leader in startups. He built SAAS and enterprise AI software from 0 to 1 and has rich experience driving go-to-market and growth strategy. He is an angel investor invested in dozens of AI startups in multiple industries. He also serves as advisors on two startups on product development and strategy. He holds a PhD degree from Stanford University in Electrical Engineering.

GPT-4 and Midjourney: Co-pilots

Dedication

We dedicate this book to the visionary thinkers and pioneers of the generative AI era. Their relentless pursuit of innovation and understanding has reshaped the landscape of product management and technology. To those who dare to dream and challenge the status quo, this work is a tribute to your spirit of exploration and creativity.

To the timeless spirit of exploration and the relentless pursuit of knowledge that dwells within us all. May this work ignite a spark of curiosity and wonder in the hearts of its readers.

Table of Contents

Foreword

Jia Li Bio: Former Founding Global Head of R&D and Co-Founder of Cloud AI/ML at Google, Former Head of Research at Snap, Advisor to UNICEF, Chief AI Fellow at Accenture, Group Lead of the Visual Computing and Learning Group at Yahoo! Labs, Co-Founder, Chief AI Officer & President at LiveX AI. Adjunct Professor at Stanford, Ph.D. in Computer Science from Stanford University, Computer Vision Foundation Industrial Advisory Board Member, IEEE Fellow, and Young Global Leader by the World Economic Forum. Li's work has been featured in the premier media outlets including Forbes, TechCrunch, CNBC, New Scientist, MIT Technology Review and more.

Foreword by Jia Li:

In the ever-evolving world of Generative AI, where each day brings new challenges and opportunities, "Reimagined: Building Products with Generative AI" by Shyvee, Caitlin and Dr. Rong emerges as a vital guide. As someone who has navigated the AI field for many years, I approach this book not only as a reader but as a fellow traveler on this journey of continuous learning and adaptation. I invite you to join me in exploring a landscape that is as challenging as it is exhilarating. This book isn't just about the awe-inspiring pace of innovation in Generative AI; it's a candid reflection on the realities of bringing these products to life.

In this book, you'll find an honest account of the complexities and rewards of developing AI-driven products. The challenges of compliance, customer education, and creating truly trustworthy products are not understated. In my own experience, understanding these nuances has been both enlightening and daunting. This book aims to share that understanding in a way that is both practical and inspiring.

For those of you who are product managers, technologists, or startup founders, this book will resonate with your experiences. It offers a deep dive into the Generative AI field, not just from a high-level overview standpoint but with hands-on guidance that is rooted in real-world application. This isn't just about the broader vision of AI; it's about the tangible steps to build something meaningful in this new era.

To the marketing leaders and those at the helm of product development, this book offers detailed guidance about what it takes to achieve product-market fit and develop effective strategies in an AI-influenced market. The tools and metrics discussed here aren't just theoretical; they are borne out of actual successes and failures.

And to the new product managers or those new to the realm of AI, this book is a sincere guide. It's about improving your skills not through lofty ideals but through practical, real-world advice and best practices. It's about understanding the opportunities and challenges in a field that is as dynamic as it is unpredictable.

Shyvee, Caitlin and Dr. Rong's perspectives are a valuable addition to the conversation around AI in product, providing guidance and inspiration for anyone looking to navigate this dynamic field, whether you're just starting out or have been in the industry for years.

As we turn the pages of "Reimagined: Building Products with Generative AI," we find ourselves equipped with the knowledge and skills to harness the power of Generative AI, transforming the way we conceive, develop, and market products. This book is a testament to the endless possibilities that await in the Generative AI Era, and an invitation to join a community of thinkers, innovators, and dreamers who are navigating this exciting yet challenging path.

Welcome to the future of product management—a future where imagination and artificial intelligence converge to create extraordinary outcomes.

Jia Li

December 2023
Palo Alto, California

Pioneering AI: Our Adventure from Curiosity to Creation

With the launch of ChatGPT on November 30, 2022, the world shifted. This groundbreaking AI technology quickly captured widespread attention overnight; AI wasn't just a topic for tech enthusiasts or sci-fi fans; it became a household discussion.

Dinner tables buzzed with debates about AI's potential and pitfalls. It felt like we were on the cusp of a new era, a sentiment echoed by AI entrepreneur, co-founder and former head of applied AI at DeepMind Mustafa Suleyman: "We are approaching a critical threshold in the history of our species. Everything is about to change."

This book began as a spark of curiosity, a need to answer a deluge of questions about AI. How could generative AI revolutionize product management? What ethical guidelines must we consider in AI product design? How can we ensure AI is accessible and trustworthy? What does the future of human-machine collaboration look like? Perhaps most fundamentally: what does it mean to be human in an age where machines can think and create?

One June evening in San Francisco, Shyvee reconnected with Caitlin and another old friend over dinner in the Potrero Hill neighborhood. The conversation, fueled by excitement and possibilities, veered towards the idea of writing a book about generative AI, leveraging generative AI itself. It was an idea too enticing to pass up.

What started as a casual dinner idea turned into a 10-hour marathon session at AGI House - an AI hacker house in Silicon Valley on a late Friday afternoon. To outline our vision for the book, we spent hours tinkering with a new tech from Omni Labs, a startup that leverages AI for book writing, founded by Jeremy Nixon, an ex-researcher from Google Brain. The initial weeks were a whirlwind of creativity and productivity, with weekends dedicated to drafting our vision with the help of AI.

As we shared our early drafts with friends in product management, the feedback was eye-opening. Juggling a full-time job, managing family obligations, and dedicating weekends to our book project began to take its toll. Considering the rapid evolution of AI, with daily updates in products, research, and companies, we pondered over the distinct insights essential for this ambitious project. We questioned how to stay focused amidst the noise and identify the core essence of our work. Which areas should we explore in-depth to provide values for the readers? Although our combined expertise was evident, the increasing demands and the dynamic nature of the AI field made us feel not quite ready, risking the derailment of our project as we struggled with burnout.

In our darkest hour, feeling on the brink of giving up, we found a lifeline. Caitlin invited a third co-author to join us, Dr. Yiwen Rong – a Stanford PhD, an AI Product Manager at a top tech FAANG company, a two-time entrepreneur, and investor. His expertise in AI and entrepreneurship breathed new life into our project.

In addition to onboarding a third co-author, to deepen her understanding of the rapidly evolving generative AI landscape with real-world examples, Shyvee initiated a special program within her PM Learning Series: interviewing leading minds at the forefront of developing cutting-edge generative AI products including Head of Robblox Studio GenAI, CEO at Uizard, VP of Product at Microsoft, Cofounder at Copy.ai, Head of Product at Synthesia, CTO and Cofounder at Forethought, Head of Product at Google Language AI, former AI Principle PM of Alexa AI, Director of Product at Instacart and more. Through these interviews, coupled with writing newsletter recaps and sharing knowledge with a robust global community of over 100,000 followers, we significantly expanded our knowledge. Together, we gained insights into the processes, technologies, and tools companies are employing to integrate generative AI into their product offerings. We explored the problem spaces these innovations aim to solve and how the new technology can impact people's lives profoundly. Additionally, we debated about the strategic challenges these companies face in pursuing sustainable growth and maintaining a defensible position in a competitive market largely in favor of the incumbents.

This book is more than just pages and words; it's a testament to our journey through the world of generative AI. It features over 150 real-world examples, 30 case studies, and 20+ frameworks. The book offers an extensive guide for integrating generative AI into product strategy and careers.

The book is divided into three parts:

Part I: Exploring the Landscape of Generative AI

The first part of the book delves into the revolutionary realm of generative AI, which has evolved from the realms of science fiction into a significant aspect of our daily lives. This section provides a primer on the AI revolution, highlighting how generative AI stands at the vanguard of numerous industry transformations. It offers a balanced view of this technology, showcasing its ability to innovate and create, while also navigating the complexities and uncharted territories of artificial imagination. Be prepared to embark on a journey into an exciting future filled with limitless possibilities and new realms to explore.

Part II: Building Generative AI Products

Part II explores building generative AI products, starting with effective customer segmentation and selection, exemplified by Synthesia's case study. It discusses the crucial balance between problem-first and tech-first approaches, employing the Jobs-to-be-Done framework, as seen in Intercom's strategy. The section emphasizes the importance of designing impactful MVPs, exploring open source and proprietary models, with insights from Neeva and BuzzFeed's experiences. It addresses the challenges in scaling, measuring success, and responsible AI development, illustrated through ChatGPT and Instacart examples. Part I concludes with a nuanced discussion of the concept of 'moats' in the generative AI industry, presenting diverse viewpoints on their necessity and effectiveness.

Part III: Navigating the Product Career in the AI Era

The final part addresses the evolving role of product managers in the age of AI. This section aims to guide product managers on adapting and thriving in this AI-empowered world, emphasizing the need for continuous learning and strategic career planning. It discusses how product managers can AI-proof their skill sets by combining emotional intelligence, creativity, and domain knowledge to design products that resonate on a human level. This section prepares product managers for the future, where they can leverage AI to innovate and collaboratively push the boundaries of what's possible in product management.

We invite you to join us on this enlightening journey. Your thoughts, critiques, and insights are not just welcomed, but essential. Writing this book has been a challenging yet rewarding endeavor, a true labor of love born from our deep passion for demystifying AI and its impact on product management and beyond. The path wasn't easy, but the rewards of sharing this knowledge are immeasurable.

AI is not just a fleeting trend—it's a transformative force that's here to stay. To keep pace with this rapid evolution, continuous learning and adaptation are key. In the spirit of this ongoing journey, we humbly ask for a small yet significant favor from you. If this book has sparked new ideas, offered valuable insights, or simply intrigued you, please take a moment to leave a review on Amazon.

Your feedback is incredibly important to us. It not only inspires us to continue creating and sharing insightful content but also helps other curious minds discover this book and embark on their own adventures in learning. Each review, each shared experience, enriches our collective understanding and contributes to a vibrant community of learners and innovators.

As you turn these pages, we hope you feel the same excitement and curiosity that drove us to create this book. Together, let's dive into the fascinating world of generative AI and explore the endless possibilities it holds for the future of product management. Welcome aboard, and let the journey begin!

Shyvee, Caitlin, and Dr. Rong

December 2023
San Francisco

Part I: Exploring the Landscape of Generative AI

1.1 - The AI Revolution: A Primer

The Artificial Intelligence (AI) Revolution has swiftly transitioned from the pages of science fiction to the fabric of our daily reality, spearheading unprecedented changes in numerous industries. At the forefront of the AI Revolution is generative AI, a technology tinged with both promise and caution. While generative AI dazzles with the potential to create and innovate, it walks a tightrope, balancing between groundbreaking advancements and the uncharted territories of artificial imagination. It's not just a technological marvel; it's a journey into the exciting, yet unpredictable, future of boundless possibilities and unexplored realms.

With a little help from our AI friends, GPT-4 and Midjourney – let's embark on a fun adventure with an AI "guest": Alan Turing, the British mathematician and computer scientist widely considered as the father of theoretical computer science and artificial intelligence. Let's take a journey to unveil the revolution of AI! GPT-4 will provide the text and Midjourney will provide the images for our exciting storyline.

> *GPT-4 Prompt: Imagine you are Alan Turing traveling through time to create a new sci-fi documentary. Educate the public about key phases in the AI Revolution, and ups and downs leading to the recent popularity of generative AI. Write it in a fun and entertaining way.*

GPT-4 (text) + Midjourney (images):

Title: Turing's Time-Travelogue: The AI Revolution
Scene: A retro-futuristic laboratory adorned with blinking lights and spinning gears. The camera zooms in to reveal Alan Turing, dressed as a nerd from the '40s and resembling a time-traveling detective. He stands beside a chalkboard filled with doodles of robots and futuristic cities.

Alan Turing as conceptualized by Midjourney, a generative AI tool

AI Alan Turing: Ah, greetings, Earthlings, time-travelers, and AI aficionados! I'm Alan Turing, and if you're wondering why I'm not locked in the 1940s, let's just say I've upgraded my travel plans. (*Presses a button, opening a time portal.*) Ready for a rollicking tour of the AI revolution?

We dive into the portal, emerging in the swinging '60s.

AI Turing: Ah, the 1960s! Picture it: The Beatles are on every radio, miniskirts are in, and here in the lab, we're betting big on machines that can think

and play chess. (*Turns to a chalkboard full of robot sketches and equations.*) You see that? That's not doodling; that's ambition! We're talking about machines that could join Mensa and never pay dues.

But here's the kicker—like a stand-up comedian who's great at open mics but bombs on the big stage, we oversold it. Yep, we promised the moon landing of AI and delivered...well, a paper airplane.

The screen transitions to a scene depicting "AI Winter" in a deserted 1970s computer research laboratory.

AI Turing: Flash-forward to the '70s—when disco was king, and AI was...well, a pauper. (*Looks around the deserted lab.*) The funding froze, the labs turned into ghost towns, and the dreams... well, they took a little nap. It was the end of season one, leaving viewers at the edge of their seats, wondering if the show would ever return.

The scene changes to a vibrant college courtyard in the 1980s.

AI Turing: But wait, there's a plot twist! Just when you thought AI was the one-hit-wonder of the science world, the '80s brought a splash of neon and a fresh wave of AI endeavors. Expert systems were the new rockstars, stealing the limelight with their cool logic and rule-based swagger. Alas, like a sitcom suffering a sophomore slump, the hype couldn't last, and AI faced another chilly reception.

Next, we are sitting inside a bustling 1990s internet café.

AI Turing: Hold on to your floppy disks, because the '90s brought the world wide web and a quiet undercurrent of AI renaissance. Yes, our quirky cast of AI visionaries were back, working behind the scenes, ready for their grand comeback. The stage was set for a new millennium, a fresh start, and boy, did they have a surprise in store for us!

Fast forward to the 2000s, 2010s, and 2020s: a time of rapid technological advancements.

AI Turing: And here we are, in the golden era of AI, where dreams are no longer just dreams, but realities unfolding before our eyes. Generative AI entered the scene, a starlet with a knack for creativity, writing poems, painting pictures, and even composing symphonies. It was the grand revival, the show everyone talked about, where machines learned to be artists, thinkers, and dreamers, taking center stage in the grand theater of innovation.

Returning to the futuristic lab, Turing steps out of the time travel machine and faces a vivid holographic projection of himself.

AI Turing: So, dear viewers, as we close this quirky, dramatic, and utterly unpredictable show of the AI Revolution, remember this is not the end, but just the beginning. Who knows what thrilling adventures await in the next season of the AI saga? Stay tuned, because the future is a script yet to be written, and with AI, the sky is not the limit, but just the beginning!

The screen fades to black, and the words "To be continued..." appear, leaving viewers with a sense of wonder and anticipation for the future.

Thank you, AI Alan Turing, for the mini time-travelogue. Get excited as we embark on this intellectual journey to grasp the full potential of generative AI!

Let's start by defining AI and debunking some of the most common perceptions about AI.

1.11 - What Is Artificial Intelligence (AI) and Generative AI?

The Oxford dictionary defines "artificial intelligence" as: "the theory and development of computer systems able to perform tasks that normally require human intelligence, such as **visual perception**, **speech recognition**, **decision-making**, and **translation between languages**"[1].

Generative AI is a transformative branch of artificial intelligence that focuses on generating new data or content that mimics existing patterns, structures, or sequences. It's like the creative artist in the AI world, capable of crafting anything from new images to intricate narratives.

To make things more interesting and inclusive, let's hear from a wide range of perspectives (from an AI):

Prompt: What is AI? Please answer from the perspectives of the following:

- A 5-year-old kid
- A 85-year-old grandma
- An iron chef at a top Michelin star restaurant
- A YouTube influencer
- A product manager at a tech company
- A construction worker
- An investment banker working on Wall Street
- A real estate broker
- A journalist writing for a top publication
- A digital marketer and brand strategist
- An up-and-coming singer at a local bar

Make your answers fun and enlightening! No more than 30 words for each persona.

GPT-4:

5-year-old kid:
AI is like a magic toy that can talk and think by itself!

85-year-old grandma:
It's like a smart radio, dear, that knows and learns!

Iron Chef at a top Michelin star restaurant:

AI's a kitchen helper who suggests recipes and remembers tastes but doesn't eat!

YouTube influencer:
AI? It's the cool tool that helps me edit videos and suggest content ideas!

Product manager at a tech company:
AI's the brain behind our apps, making them smarter and user-friendly.

Construction worker:
AI's that fancy hat which warns me about unsafe zones on site.

Investment banker working on Wall Street:
AI's the whiz-kid assistant that predicts market trends and gives stock tips.

Real estate broker:
AI helps me match dream homes with the right buyers faster!

Journalist writing for a top publication:
AI's my research buddy, quickly pulling up facts and trends.

Digital marketer and brand strategist:
AI analyzes data, sometimes making campaigns hit-or-miss. It's a tool, not magic.

Up-and-coming singer at a local bar:
AI might push my song or drown it in the sea of tunes. It's unpredictable.

1.12 - What Have You Been Getting Wrong About AI?

Here are some widely-held misconceptions about AI:

1. All AI is the same.
Many people use the term "AI" to refer to a wide range of technologies, from simple rule-based systems to complex deep learning networks, assuming that all

AI technologies function similarly. In reality, there is a vast range of AI technologies, each with its own strengths and weaknesses, that can be applied to solve various problems.

The below diagram illustrates the relationships of various AI disciplines to each other and to AI as a whole. See the Appendix to learn more about these key concepts in AI.

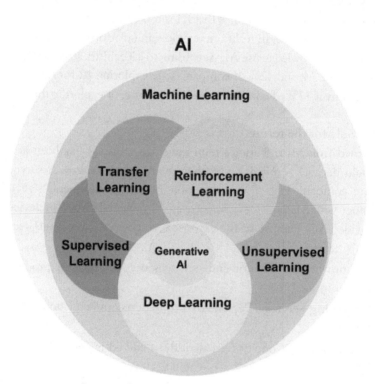

Copyright ©2024 by Authors of Reimagined.

2. AI can "understand" humans in the way other humans do.

While AI can process and analyze large amounts of data to identify patterns and make predictions, it doesn't "understand" information in the human sense. AI lacks the innate human ability to comprehend nuances, context, and emotions fully.

3. AI can replace all human jobs.

AI can certainly automate many tasks, but there are still many jobs that require a level of creativity, empathy, and understanding that AI cannot replicate. Moreover, AI can create new opportunities and jobs by taking over repetitive tasks and allowing humans to focus on more complex and creative endeavors. We talk about <u>career development in the AI-empowered era</u> in Part III.

4. AI is only for tech-savvy individuals or large corporations.

A very common misconception is that AI is too complex for the average person or small business to use. In reality, many tools and platforms help people without deep technical knowledge use AI. According to a TechJury article, 35% of global companies use AI, and 42% of companies are exploring AI for future use. That means that over 77% of companies are either using, or exploring the use of, AI[2].

5. AI is only for the future.

We learned from Alan Turing's mini-travelogue that AI has been around for over eight decades – much longer than many people realize. A survey by the software company Pega found that only 33% of consumers *think* they use technology with AI, but 77% *actually* use an AI-powered service or device. Below we provide just a glimpse of the expansive reach of AI in various sectors and daily activities[3]:

- **E-Commerce**: Recommendation engines to suggest products (e.g., Amazon).
- **Streaming Services**: Personalized content recommendations (e.g., Netflix, Spotify).
- **Social Media:** Content feed algorithms and ad targeting (e.g., Facebook, TikTok).
- **Voice Assistants:** Virtual helpers answering queries (e.g., Siri, Alexa).
- **Email:** Spam filters and categorization (e.g., Gmail).
- **Photography:** Image recognition and auto-tagging (e.g., Google Photos).
- **Navigation:** Real-time traffic predictions (e.g., Waze, Google Maps).
- **Banking:** Fraud detection and credit score predictions (e.g., Citi, Credit Karma).
- **Healthcare:** Predictive health monitoring (e.g., Whoop, Apple Watch).

- **Retail:** Virtual try-ons and style suggestions (e.g., Sephora's virtual artist).
- **Gaming:** Adaptive game difficulty and Non-Player Character (NPC) behaviors.
- **Customer Support:** Chatbots answering common queries (e.g., website support).
- **Home Automation:** Smart thermostats adapting to user behaviors (e.g., Nest).
- **Agriculture:** Predictive analytics for crop yields (e.g., Cropin, Blue River Technology).
- **Dating Apps:** Matchmaking algorithms (e.g., Tinder, Hinge).
- **Transport:** Ride-sharing price estimations and route optimizations (e.g., Uber, Lyft).
- **Shopping:** Virtual assistants helping with online shopping queries (e.g., eBay's ShopBot).
- **Language:** Real-time translation tools (e.g., Google Translate).
- **Education:** Personalized content suggestions and learning paths (e.g., Khan Academy, Coursera).
- **Real Estate:** Property suggestions based on user preferences (e.g., Zillow, Redfin).

1.13 - Why Is AI an Old Phenomenon?

As Alan Turing's mini-travelogue illustrates, the odyssey of AI is not merely a technical journey, but a tale of ambition, hope, and unexpected pitfalls. From humble beginnings in quiet labs to global notoriety and skepticism, AI's evolution reflects humanity's relentless pursuit to comprehend and control the intricate mechanisms of intelligence itself. And as with all great tales, to truly grasp its essence, one must traverse the past to appreciate the peaks and troughs of an innovation that dared to replicate the very essence of human thought. Let's trace the history that brought AI from concept to reality.

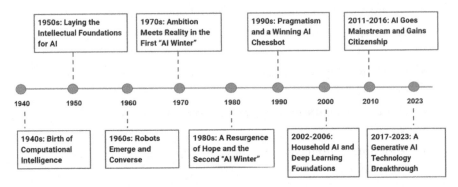

AI's origin story takes a nonlinear path with fits and starts and even two AI "winters" characterized by a lack of progress, funding, and interest[4].

1940s: Birth of Computational Intelligence

The Second World War catalyzed the embryonic stages of AI. In 1942, the Enigma code was decrypted using Alan Turing's Bombe machine, marking AI's first real-world application. Turing's genius didn't stop there. By 1948, he introduced the world to Turochamp, the first computer program trained to play chess.

Wartime picture of a Bletchley Park Bombe[5].

1950s: Laying the Intellectual Foundations for AI

In 1950, Turing introduced the famous "Turing Test" in his paper, "Computing Machinery and Intelligence." His work was inspired by that of Walter Pitts and Warren McCulloch, which introduced the idea of networks of idealized artificial brain cells, or neurons, called "neural networks." In 1951, Marvin Minsky and Dean Edmonds built the first neural net machine, called SNARC (short for "Stochastic Neural-Analog Reinforcement Computer"), which marked a significant step in AI[6]. In the summer of 1956, AI was born at a workshop at Dartmouth University in New Hampshire. Attendees, luminaries in their fields, envisioned machines rivaling human intelligence within a generation. The computer scientists' optimism attracted significant attention and funding, setting the stage for decades of future research.

1956 Dartmouth Conference: The Founding Fathers of AI

The above visionaries attended the 1956 Dartmouth Conference, which is considered to be the birthplace of AI. These founding fathers of AI talked about the future of computers that could model human intelligence[7].

1960s: Robots Emerge and Converse

The 1960s were a period of tangible AI innovations. 1961 welcomed Unimate, the first industrial robot, onto a General Motors assembly line. By 1964, the first chatbot, ELIZA, was developed at MIT to simulate psychotherapy sessions. The middle of the decade, 1966, introduced Shakey, the first general-purpose AI-powered robot capable of reasoning through its actions[8]. Specifically, Shakey was able to move around on its own and navigate through various rooms. Its functionality included "a TV camera, a range finder, radio communications, and a set of drive wheels controlled with stepping motors"[9].

Unimate pouring coffee for a woman at Biltmore Hotel, 1967[10].

```
Welcome to
                 EEEEEE  LL      IIII  ZZZZZZ   AAAAA
                 EE      LL       II      ZZ   AA   AA
                 EEEEE   LL       II     ZZZ   AAAAAAA
                 EE      LL       II    ZZ     AA   AA
                 EEEEEE  LLLLLL  IIII  ZZZZZZ  AA   AA

   Eliza is a mock Rogerian psychotherapist.
   The original program was described by Joseph Weizenbaum in 1966.
   This implementation by Norbert Landsteiner 2005.


ELIZA: Is something troubling you ?
YOU:   Men are all alike.
ELIZA: What is the connection, do you suppose ?
YOU:   They're always bugging us about something or other.
ELIZA: Can you think of a specific example ?
YOU:   Well, my boyfriend made me come here.
ELIZA: Is it important to you that your boyfriend made you come here ?
YOU:   He says I'm depressed much of the time.
ELIZA: I am sorry to hear that you are depressed.
YOU:   It's true. I am unhappy.
ELIZA: Can you explain what made you unhappy ?
YOU:
```

The first chatbot, ELIZA, was developed at MIT starting in 1964, and released in 1966[11].

1970s: Ambition Meets Reality in the First "AI Winter"

The soaring ambitions of the 1950s and 1960s tackled the intricate challenges of replicating human cognition. By the mid-1970s, however, limitations in computing power and data led to disappointing performance of AI, creating the first "AI winter." People lost hope that AI could become the reality that they had envisioned. The British and United States governments, including US agencies such as National Research Council (NRC) and Defense Advanced Projects Research Agency (DARPA), reduced funding in AI projects, skeptical that AI could ever deliver on its transformative promises.

1980s: A Resurgence of Hope and the Second "AI Winter"

The 1980s heralded renewed enthusiasm in AI that was driven by international initiatives and innovations and adoptions of expert systems. Two researchers with interests at the intersection of cognitive science and computer science, Geoffrey Hinton and David Rumelhart, popularized "backpropagation" for training neural networks and reinvigorated the exploration of artificial neural networks. Japan, in particular, emerged as a major player with ambitious goals, with $850 million funding for the fifth-generation computer project. Yet, as the '80s ended, another wave of disillusionment loomed, marking the onset of a second "AI winter." Hundreds of AI companies were acquired or shut down.

1990s: Pragmatism and a Winning AI Chessbot

The next significant milestone came in 1993 when Vernor Vinge published "The Coming Technological Singularity," warning of a future where super AI could mark the end of the human era. 1995 introduced the chatbot A.L.I.C.E, which was a significant evolution from ELIZA. An AI accomplishment that shocked the world happened in 1997, when IBM's Deep Blue, a computer, defeated the world chess champion, Garry Kasparov. This AI victory prompted the world to wonder about AI's potential to outsmart humanity.

Deep Blue was a computer developed by IBM resembling this one. Deep Blue defeated chess world champion Garry Kasparov in May 1997, making it the first computer to win a match against a world champion[12].

2002-2006: Household AI and Deep Learning Foundations

A company called iRobot got its start making robots to go anywhere from the battlefield to the living room. The Roomba, iRobot's AI vacuum launched in 2002, made AI a household name – quite literally. The early 2000s also saw Geoffrey Hinton lay the groundwork for deep learning in "Learning Multiple Layers of Representation," setting the stage for a new era of AI capabilities.

2011-2016: AI Goes Mainstream and Gains Citizenship

The early 2010s were an explosive period for AI. IBM's Watson stunned audiences by winning Jeopardy! in 2011, the same year Siri, Apple's smart device virtual assistant, was introduced. By 2016, Sophia, a humanoid robot, was granted Saudi Arabian citizenship, while Google DeepMind's AlphaGo displayed its prowess in the strategic game of Go.

Sophia, a humanoid robot from Hanson Robotics Ltd., speaks at the AI for GOOD Global Summit, ITU, Geneva, Switzerland, 7 - 9 June, 2017[13].

2017-2023: A Generative AI Technology Breakthrough

In 2017, Google's transformer models became foundational for AI leading to OpenAI's Generative Pre-Trained Transformer, more popularly known by its acronym, GPT, by 2020. According to Statista, ChatGPT reached one million users within five days after its release in 2022. While it's true that Threads from Meta also reached 1 million users in a few hours, many users came from Meta's other products such as Facebook and Instagram, rather than organically as with ChatGPT, so we don't view that as an apples-to-apples comparison. As the current decade unfolds, AI research has become a global endeavor, with over 72 active Artificial General Intelligence (AGI) R&D projects spanning 37 countries by 2020. The burgeoning field of open-sourced learning hints at a future where AI systems could continuously innovate. Yet, as AI's capabilities grow, so do the ethical considerations surrounding its application[14]. We will delve into AI ethics and how to build responsible AI products in section 2.25.

ChatGPT's milestone of reaching one million users in five days was the fastest ever. Image from Statista[15].

1.2 - The Catalysts and Precursors of Generative AI

1.21 - Why Is Now the Right Time for Generative AI?

"AI is one of the most important things humanity is working on; it is more profound than electricity or fire." declares Sundar Pichai, CEO of Google. He's not alone in this sentiment. Bill Gates identifies generative AI as one of the two most revolutionary technologies he's encountered, equating its significance to the advent of the personal computer and the Internet.

The Democratization of AI
What used to be a conversation locked within academia and elite tech circles has been democratized. OpenAI has been pivotal in this transformation. With tools like ChatGPT, they've made cutting-edge AI technology accessible, inviting the global public to partake in a conversation about the future of AI. Even an 85-year-old grandma can talk to an AI chatbot for companionship without much technical knowhow.

Technological Confluence
Several factors have contributed to the rise of generative AI:
1. **Data Availability**: We live in a world flush with data. This rich sea of information serves as the foundational bedrock for training and fine-tuning generative AI models.
2. **Hardware Evolution**: The advent of specialized hardware like graphics processing units (GPUs) and tensor processing units (TPUs) has turbocharged the training speed of intricate AI models.
3. **Model Architecture**: A driving force behind the evolution of generative AI is the rise of advanced neural networks, most notably deep learning architectures like generative adversarial networks (GANs) and transformers. GANs involve a dueling pair of neural networks—a generator and a discriminator—working in tandem to improve data generation capabilities. Coupled with transformers, whose attention mechanisms were popularized by the groundbreaking 2017 paper "Attention is All You Need" by Google Research, these technologies form

the foundational bedrock for generative AI. Altogether, these advances serve as both the canvas and the brushes, enabling the generation of an incredibly broad and nuanced spectrum of outputs. However, the scope of innovation extends beyond just language; newer techniques like Stable Diffusion and neural radiance fields (NeRFs) are expanding the possibilities even further. Stable Diffusion adds a layer of randomness to data generation, which is then methodically reversed, while NeRFs provide a gateway for turning 2D images into detailed 3D models.

An Analogy: The Culinary Art of AI

Developing an AI model is akin to cooking a gourmet meal. You need the right ingredients (data), a recipe (model architecture and hyperparameters), culinary skills (organizational capabilities), and state-of-the-art kitchen appliances (GPUs), but even the perfect dish is incomplete without its final garnish. In the world of AI, the "killer app" serves this role—in other words, software that is so good that it becomes a flagship or cornerstone technology. In AI, the "killer app" could be a chatbot that speaks like a human or a recommendation engine that knows what you want before you do.

So, why is now the best time to dive into generative AI? The answer is that we're at a confluence of data availability, hardware capability, and algorithmic innovation, catalyzed by a surge of global interest. The table is set, the kitchen is ready, and the world is hungry for what generative AI has to offer. Now is the time to serve this transformative feast to the world, and everyone is invited to partake.

1.22 - Is Generative AI Really the Future?

While it's tempting to dismiss the current buzz around AI as just another fleeting trend, given its long and hype-prone history, this time is different. Companies in the AI sector (like OpenAI's ChatGPT) have shown an unparalleled level of consumer engagement and rapid adoption. Since capturing public attention in the latter half of 2022, generative AI has been the catalyst for some of the most meteoric rises in companies, products, and initiatives ever witnessed in the tech world.

The Promise of Generative AI: Is It the Next Big Thing?

The Superpowers of Generative AI

What makes Generative AI particularly interesting is the various new value propositions that could bring step-change improvements to existing product experience. We refer to these transformative capabilities as **"Generative AI Superpowers."**

1. CREATIVE CONTENT GENERATION
Create diverse, multimodal content and new knowledge

2. DATA GENERATION SIMULATION, & PREDICTION
Simulate scenarios, forecast outcomes, and strategize plans

3. TASK GENERATION, PLANNING, & EXECUTION
Plan execute, and refine intricate tasks

4. ADAPTIVE PERSONALIZATION
Learn from feedback to tailor unique experiences

5. REAL-TIME INTERACTIVITY
Analyze and react to real-time data and user inputs

6. ACCESS DEMOCRATIZATION
Broaden access to knowledge and skills

Copyright ©2024 by Authors of Reimagined.

1. **Creative Content Generation:** Generative AI excels in creating diverse content—be it text, image, audio, video, code, or 3D design. This superpower is revolutionizing everything from automated marketing to website design, unlocking new realms of creative possibility.
2. **Data Generation for Simulation and Prediction:** In industries where data is scarce, generative AI creates synthetic datasets, filling the gaps in our knowledge. It not only aids in decision-making, but also forecasts future scenarios, giving businesses a crystal ball for market trends.

3. **Task Generation, Planning, and Execution:** From planning intricate tasks to executing them efficiently, generative AI acts as a taskmaster. It even learns from past performances, optimizing each step in the problem-solving process and making continuous improvement second-nature.

4. **Adaptive Personalization:** Generative AI stands out due to its dynamic personalization ability, driven by advanced learning mechanisms. It tailors experiences to individual needs, be it recommending products or shaping educational paths.

5. **Real-Time Interactivity:** Real-time AI is always in the moment. Whether rerouting traffic in real-time or adapting a video game to a player's actions, its ability to analyze and update in real-time sets a new standard for system responsiveness.

6. **Democratizing Expertise**: Generative AI levels the playing field, making expert-level resources accessible to novices. It's the driving force behind a growing suite of tools for non-technical users.

These superpowers of generative AI are not restricted to specific industries but are applicable across diverse sectors. They provide a flexible framework that can be used to understand and evaluate the potential impact of AI in various contexts. As we continue to explore and innovate with generative AI, these superpowers will be instrumental in shaping a future where AI is an integral part of our lives. We will look at the various generative AI applications in section 1.4.

The Market Potential

According to a June 1, 2023 Bloomberg Intelligence report, the generative AI landscape is on the brink of a transformative expansion. The market, valued at $40 billion in 2022, is projected to see an explosive growth to a staggering $1.3 trillion within a decade, at a CAGR of 42%[16].

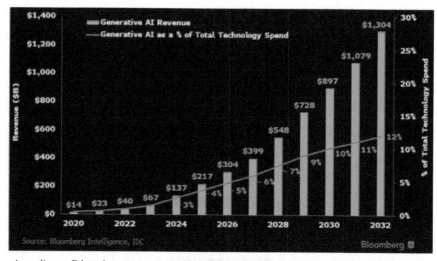

According to Bloomberg, generative AI will become a $1.3 trillion market by 2032. Image from Bloomberg LP[16].

Sustained Market Adoption and Economic Efficiency

ChatGPT's rapid ascent to one million users in five days and being able to sustain 230 million monthly active users in just six months is a testimony to generative AI's market potential. This growth eclipsed even that of Facebook in its early days; it took Facebook five years after initial launch to achieve a comparable 197 million monthly active users[17].

ChatGPT is not alone in its expansive popularity. Companies like Midjourney and Character.AI have also seen explosive growth: Midjourney's Discord server swelled to nearly 15 million members in less than a year, and Character.AI garnered 18 million monthly unique web visitors and exceptionally high user engagement just nine months post-launch. Even newcomers like Janitor AI amassed a million users within weeks of launching[17].

Unlike traditional AI, generative AI doesn't rely on strict notions of 'correctness,' making it ideal for areas like content creation and companionship. Its versatility enables it to disrupt multiple sectors, from entertainment to professional services like therapy and legal advice, thus serving as a launchpad for entirely new industries.

Economically, generative AI offers astounding efficiencies, performing tasks at a fraction of the time and cost of human alternatives. For instance, creating an image costs as little as $0.001 and takes seconds, while a human artist might require hundreds of dollars and hours. These economic advantages extend to high-wage sectors, offering not just incremental benefits but transformative shifts in cost structures.

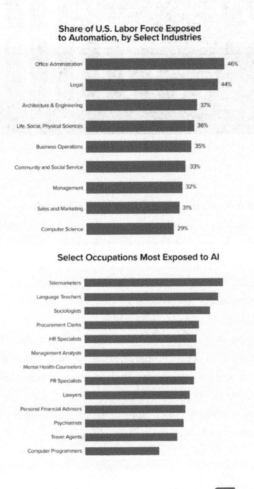

Share of U.S. Labor Force Exposed to Automation, by Select Industries

Industry	%
Office Administration	46%
Legal	44%
Architecture & Engineering	37%
Life, Social, Physical Sciences	36%
Business Operations	35%
Community and Social Service	33%
Management	32%
Sales and Marketing	31%
Computer Science	29%

Select Occupations Most Exposed to AI

Telemarketers
Language Teachers
Sociologists
Procurement Clerks
HR Specialists
Management Analysts
Mental Health Counselors
PR Specialists
Lawyers
Personal Financial Advisors
Psychiatrists
Travel Agents
Computer Programmers

Source: Goldman Sachs, March 2023, Felten, Raj, and Seamans (March 2023)

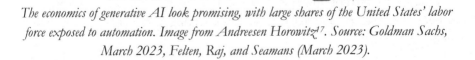

The economics of generative AI look promising, with large shares of the United States' labor force exposed to automation. Image from Andreesen Horowitz[17]. Source: Goldman Sachs, March 2023, Felten, Raj, and Seamans (March 2023).

These revolutionary economics are backed by real revenue. ChatGPT alone is approaching an annual run rate of $500 million, and other generative AI ventures are similarly scaling their revenues, often outpacing their operational costs[17]. This rapid financial scaling and user willingness to pay suggest more than just economic viability; they signal a paradigm shift. Much like the microchip and the Internet, generative AI is emerging as a groundbreaking technology that's poised to redefine industries and usher in a new epoch in computing and human-machine interaction.

The Skepticism: Is Generative AI Overhyped?

While the buzz around generative AI is palpable, it's crucial to temper enthusiasm with a dose of skepticism. For starters, much of the technology still resides in the "demo" or "playground" phase, lacking a proven path to stable commercial viability. Then there's the risk of "hallucinations"—inaccurate or fabricated information generated by AI. These hallucinations not only raise questions about the reliability of AI, but also introduce a host of ethical and legal concerns, such as misinformation and data integrity. Scaling AI technologies introduces another set of challenges. As demand for computing resources soars, we're looking at potential shortages that could stifle development and deployment.

Furthermore, the current landscape of generative AI applications resembles a sea of sameness, as many technologies are built on similar underlying models (also known as "GPT wrappers"). This lack of differentiation makes it hard for companies to build sustainable competitive moats. Therefore, while the initial signs are promising, a 'wait and see' approach might be prudent. We're still in the early days of AI development, and the path forward is fraught with challenges and limitations that we'll explore in greater depth in section 1.5.

1.23 - Why Are We Still Early in the AI Evolution?

In the grand scheme of AI evolution, we're still very much in the "childhood" phase, primarily because most of the AI models we interact with are specialized, but lacking the general cognitive capabilities that define human intelligence.

Where are we, exactly? Let's recap some of the AI technologies that currently exist:

- **Narrow AI (Weak or Traditional AI)**: This is the most prevalent type of AI today. Examples include Siri, Roomba, and chatbots like ELIZA and A.L.I.C.E. These models excel at specific tasks, but can't think beyond their programming. Their capabilities are narrow, defined by the data they were trained on and the rules they were programmed to follow.

- **Reactive Machines**: Think of IBM's Deep Blue, the chess-playing AI. These machines don't have a "memory" of past actions; they analyze situations in real-time and make decisions based on that instant analysis.

- **Limited Memory**: These AI systems, like self-driving cars, do use past experiences to make future decisions, but their learning is still constrained. They can't generalize this learning to solve new, unforeseen problems.

- **Generative AI (we are here)**: This is the rising star of the AI world, capable of creating new content such as text, images, or even more abstract constructs like computer code. It shows promise in offering more dynamic and adaptable solutions, but is still in the nascent stages of commercial viability.

Where can we go? Here are some potential future goals of AI:

- **Artificial General Intelligence (AGI) or Strong AI**: This is the holy grail of AI—machines that can think, reason, and learn across any intellectual activity. We've had glimpses of this potential with Sophia and AlphaGo, but we're still far from realizing it.

- **Artificial Super Intelligence (ASI)**: This theoretical future state of AI would not only emulate human intelligence but surpass it in all aspects, including creativity and social interactions.

- **Theory of Mind and Self-Aware AI**: These are theoretical constructs where AI would understand emotions, beliefs, and even have self-awareness. We're nowhere close to achieving this level of sophistication and there are many debates among scholars if AI will ever achieve this level of cognition.

Why are we still so early in the AI evolution? The reason is that we're mostly dealing with specialized or 'narrow' AIs that excel at one thing, but falter when asked to generalize their understanding to broader contexts. Even Generative AI, which is more versatile than its predecessors, faces challenges in accuracy, control, and ethical considerations. The 'future goals' of AI—AGI, ASI, and self-aware AI—remain largely theoretical, underscoring just how much room there is for growth, improvement, and discovery in this fascinating field.

1.3 - Generative AI Market Structure and Tech Stack

<u>Note</u>: This chapter delves into the more technical aspects of building generative AI. While it offers a deeper understanding of the intricacies involved, it may be more detailed than some readers require. Feel free to engage with this chapter to enhance your technical knowledge, or skip ahead if you prefer to focus on the broader concepts and applications of generative AI.

1.31 - How's the Generative AI Scene Structured and Who's Winning?

Before we talk about leaders in the AI space, let's talk about the way this space is structured. Three layers characterize the generative AI tech stack: the foundation layer, the tool layer, and the application layer. In this section, we'll talk about each of these layers in detail.

Three Layers of Generative AI Companies

Layer 3: Application Layer

Layer 2: Tool Layer

Layer 1: Foundation Layer

Layer 1: Foundation Layer
The Foundation Layer is the bedrock infrastructure upon which everything else stands. Here, we find essential components like hardware and cloud platforms, data sources and foundational AI models - the infrastructure, blueprint, and raw materials needed to begin construction. These are the essential components for AI technology, both in terms of hardware and software.
- **Hardware:** GPUs (Nvidia), TPUs (Google)
- **Cloud Platform:** Amazon Web Services (AWS), Google Cloud, and Azure

- **Data:** Open source data library and proprietary data
- **Foundational Models:** Closed Source Foundation Models (OpenAI, Anthropic, Cohere, etc.), Open Source Foundation Models (Stable Diffusion, GPT-J, FLAN T-5, and Llama)

Layer 2: Tool Layer

With the right tools, an artist can paint a beautiful landscape. The same is true with AI. The Tool Layer of generative AI is where the architects and engineers have their arsenal of tools at their disposal, empowering developers to realize their visions, tapping into domain expertise without delving deep into the AI infrastructure intricacies.

- **Data-Focused AI Tools** facilitate data preparation, labeling, storage, and indexing (such as vector database), data management (versioning, governance), and more.
- **Model-Focused AI Tools** empower model selection, training, turning, evaluation, validation, simulation, monitoring, etc.
- **Developer Tools and Frameworks** connect the rapidly-evolving AI development to streamline the application development process and enhance iterations.

Layer 3: Application Layer

The Application Layers consist of user-facing products that integrate generative AI technology. It can be categorized by horizontal applications (by modality and function) or vertical application (by industry).

- **By Modality:** text, audio, image, video, 3D, code, multimodal, actions, robotics, agent, etc
- **By Function:** sales & customer support, design, search, security, productivity tools, etc
- **By Industry:** education, consumer, entertainment, legal, finance, healthcare, mobility, etc

The three layers – foundation, application, and tool – come together to create a powerful and versatile generative AI landscape. Below is a generative AI landscape map which shows a conceptual representation of these layers in action to develop generative AI products.

Generative AI Landscape Map

Application Layer	By Corporate Function					End to End Apps End-user facing apps with proprietary models and data
	Engineering	HR	Finance	Safety/ Compliance	Sales Marketing	
	Design	Productivity	R&D	Search		
	By Industry					
	Education	Consumer	Entertainment	Legal	Finance	
	Healthcare	Industrial	Mobility	Proptech	Greentech	
	By Modality					
	Text	Code	Audio/Voice	Video	3D/Animation	
	Music	Multimodal	Logics, Action	Agent	Robotics	

Tool Layer	Developer Tools and Frameworks
	Data Focused Tools Data Prep, Labeling, Storage & Indexing (e.g. vector database), and Data Management (versioning, governance)
	Model Focused Tools Feature Engineering, Model Selection, Training, Tuning, Model Evaluation, Validation, Simulation, Model Monitoring & Observability, Model Hubs

Foundation Layer	Closed Source Foundation Model	Open Source Foundation Model
	Data	
	Cloud Platform: AWS, Azure, etc	
	Compute Hardware: GPUs (Nvidia), TPUs (Google)	

We (Authors of Reimagined) created the Generative AI Landscape Map using our own understanding of the landscape; it is also informed by an Andreesen Horowitz article titled "Who Owns the Generative AI Platform?"[18] as well as a Generative AI market map produced by Kelvin Mu from Translink Capital[19].

1.32 - Why Should I Care About the Generative AI Tech Stack?

A tech stack encompasses the collection of technologies, frameworks, tools and infrastructures employed in the creation and rollout of software programs.

Product managers should develop a general understanding of the generative AI tech stack. Here are just a few reasons why:

- **Informed Decision-Making:** Understanding the tech stack allows product managers to choose the right tools and technologies, aligning development with project goals.
- **Adaptability and Futureproofing:** Knowledge of the tech stack ensures flexibility in an evolving landscape, making it easier to adapt to new technologies and trends.
- **Strategic Innovation:** A grasp of the underlying technologies helps product managers identify new opportunities and applications for generative AI, fostering innovation.

What's in the Generative AI Tech Stack?

Image on the next page is an overview of the foundational elements in the tech stack for generative AI from Madrona's Generative AI Stacks Landscape Map as of June 2023.

1. Application Frameworks: The Blueprint for Building AI Solutions
Application frameworks are pre-built structures that provide a standardized way to develop software applications. In the context of generative AI, frameworks like LangChain and Fixie offer a set of tools and protocols that streamline and accelerate the development process. Product managers find these frameworks invaluable because they allow for quicker iterations and go-to-market strategies. They facilitate the creation of a variety of applications, from content generation to semantic systems, without requiring a deep technical understanding of the infrastructure level. This means product managers can focus more on customer needs and innovation, making these frameworks a cornerstone in the effective development and deployment of generative AI products.

2. Models: The Brain of Generative AI
Foundation models (FMs) serve as the brain of generative AI applications, capable of human-like reasoning. These models come in various forms, each with its own set of attributes and associated output quality, cost, and latency.

Madrona's conceptualization of the generative AI stack. Image from Madrona[20].

Developers have a buffet of choices—close-source, proprietary models or a growing array of open-source options. There are three types of FMs: general, specific, and hyperlocal AI models. We define each below.

- **General AI models** are versatile engines that can perform a broad range of tasks from vendors like OpenAI and Anthropic
- **Specific AI models** are tailored for specialized tasks, such as an AI model trained specifically to create compelling e-commerce product descriptions.
- **Hyperlocal AI models** use specialized, often proprietary, data to produce extremely accurate and tailored outputs. A hyperlocal AI model trained on a company's internal financial and operational data can generate real-time business forecasts that are specifically tailored to the company's unique market conditions and business practices.

Product managers must understand the intricacies of FMs for selecting or combining the right models to align with the product's goals, whether it's low-latency performance or high-quality output. In Part II section 2.22, we will talk about factors to consider when choosing which types of FMs to build and how to navigate the open source v. proprietary LLM continuum.

Additionally, hosting options have diversified. Advancements from companies like OctoML allow for deployment on edge devices and browsers, providing more flexibility in setting privacy, security, and cost requirements. This broadens the strategic choices available in the product development process, making a deep understanding of FMs indispensable.

3. Data: The Fuel that Powers AI Engines

Data loaders and vector databases are key components in the generative AI stack that empower smarter and more functional models. For product managers, understanding these tools is pivotal. Data loaders allow developers to pull in diverse types of data—be it structured from databases or unstructured like PDFs or PowerPoints. For product managers, understanding data loaders means appreciating how the right data sources can shape the AI's outputs for tasks like personalized content generation or semantic search. Vector databases come into play when developers want to search through unstructured data effectively. They take data, convert it into a format that the AI can understand (known as embeddings), and store it for quick retrieval. Knowing how to leverage these can guide product managers in architecting solutions that are both efficient and scalable.

Context windows add another layer of sophistication by enabling personalized model outputs without altering the AI model itself. Platforms such as LangChain and LlamaIndex offer easy ways to include this tailored information in the AI's workflow, without modifying the underlying technology.

4. The Evaluation Platform: The Testing Ground for AI Performance

Optimizing a Language Learning Model or LLM requires an evaluation platform. Prompt Engineering tools allow for prompt iteration across various models without deep technical expertise.Experimentation tools like Statsig enable machine learning (ML) engineers to experiment with prompts, hyperparameters, and fine-tuning. They also help evaluate model performance in production, circumventing the many shortfalls of offline evaluation in a staging environment. Finally, Observability platforms such as WhyLabs' LangKit offer ongoing insights into model output quality, protection against misuse, and checks for ethical AI practices. Mastering these tools ensures better project management and risk assessment.

5. Deployment: The Launchpad for Real-World Applications

Developers have the option to either self-host using platforms like Gradio or use third-party services. Fixie stands out as a solution for creating, sharing, and deploying AI agents in production environments.

As we've seen, having a contextual understanding of the generative AI tech stack—from application frameworks to deployment options—empowers product managers to make smarter decisions, adapt to evolving tech trends, and drive innovation[21,22]. Stay tuned to explore how these foundational elements bring various generative AI applications to life.

1.4 - Generative AI Applications

1.41 - What Industries Are Being Revolutionized by Generative AI?

According to CB Insights' AI 100 for 2023, 100 startups have shown remarkable promise across various sectors, collectively raising a staggering $22 billion since 2019. Microsoft's $13 billion investment in OpenAI is a notable highlight. Among these pioneering companies, there's a significant focus on generative applications, especially in the media and entertainment sector. Companies like Character.ai, Descript, and Runway are redefining fields from chatbot interactions to video editing (more on this later).

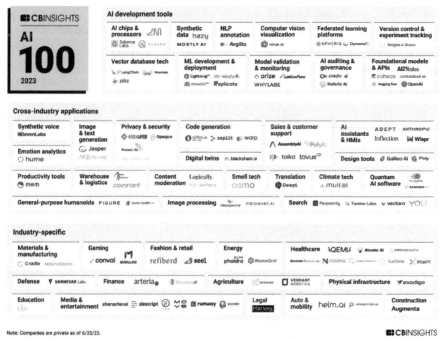

A market map for generative AI startups. Image from CBInsights[23].

Zooming In: The Application Layer

As we dive into the universe of generative AI technology, our spotlight focuses on the user-facing Application Layer. That's where AI fuses with real-world needs to create disruptive products – much like how GPS technology paved the way for game-changers like Uber and Google Maps. Product managers stand at this intersection, navigating the complexity of user needs and data streams. In this intersection of ideas and reality, AI becomes more than just a tool—it emerges as a potent enabler and a synergistic collaborator. AI technologies unlock new pathways for automation, efficiency, and optimized productivity, giving product managers a broader canvas on which to paint their visions. When viewed as collaborators, AI technologies don't merely assist, but co-create, empowering product managers to design solutions that are finely tuned to both the empirical data and the emotional contours of user needs. In its dual role as helpers and co-creators, AI harmoniously fuses with human insight, allowing product managers to not just solve problems, but to architect experiences that deeply resonate with users. Let's look at how generative AI apps are shaping how we live, work, and interact.

Note: AI and generative AI are rapidly evolving fields. The examples we've highlighted are likely to change as they face real-world market tests and as technology progresses.

Content Creation

Text Generation: Starting from text generation, leading the charge is OpenAI's ChatGPT, a versatile model adept at generating text in conversational or formal formats. The space is rapidly diversifying, with specialized offerings targeting niche areas. For instance, products like Sudowrite[24] and Verb.ai[25] have emerged to support creative writing endeavors, assisting with everything from plot brainstorming to nuanced storytelling.

Sudowrite has an AI Canvas that can generate alternate plot points, character secrets, and plot twists with the author.

The evolution of AI in text generation isn't limited to personal creativity. In the realm of social media and marketing, platforms like Copy.ai[26] harness generative AI to craft compelling marketing copy, blog posts, and product descriptions within seconds, vastly accelerating the content creation process. For sales teams, AI-powered tools like Lavender[27] aid in crafting effective outreach emails, boosting response rates and fostering customer engagement. Analytical uses have also flourished; tools like Viable[28] and Otter[29] leverage AI to process customer feedback and transcribe meetings, generating key takeaways and summaries that can inform strategic decisions.

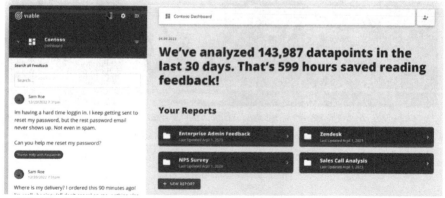

Viable allows companies to uncover powerful insights by analyzing sales call transcripts, customer support tickets, market research, employee surveys, and more.

Visual Media Generation: AI can also be used to aid in the creation of images, videos, characters, animations, and design in the visual media realm. When it comes to visual media use cases, generative AI fundamentally alters how we interact with art, video, design, and audio. At the helm of the visual media AI generation space are products like Midjourney[30] and Stable Diffusion[31], which generate unique art pieces from simple text prompts and styles, democratizing the artistic process.

Adobe Firefly[32] and Lightricks[33] redefine image manipulation, allowing users to remix, upscale, animate, and enhance images in novel ways.

In video creation, solutions like Descript[34] and RunwayML[35] are broadening their horizons beyond editing to create content from scratch. For instance, RunwayML transforms images, video clips, or text prompts into compelling film pieces. Products like Synthesis AI[36] and BHuman[37] even allow for creating engaging presentation videos or personalizing sales videos, elevating the audience's viewing experience.

To further illustrate the example, we used Midjourney to create "Unifly", a creative blend of butterfly and unicorn symbolism, embodies transformation and magic. This concept can inspire a variety of creative designs, including toys, whimsical decor, jewelry, concert stages, and architectural sculptures.

In the realm of personalized avatars, Lensa[38] enables users to train their own AI models, resulting in bespoke AI avatars. For businesses, Soul Machines[39] offers

AI "talking heads" to communicate with customers or staff more effectively. Companies like Kinetix[40] and Flawless AI[41] are innovating in the field of motion capture animation and visual dubbing, taking animation and lip-sync to new heights.

In the audio space, companies like Boomy[42] and Riffusion[43] generate music from scratch, while Resemble[44] creates studio-quality voiceovers. Furthermore, Metaphysic[45] pioneers the creation of hyper-real deepfake videos of artists and celebrities, pointing to the potentially transformative (and ethically complex) future of entertainment.

Metaphysic, a generative video AI application, can bring Elvis Presley "back to life."

Given that AI technology is still in its nascent phase, many tools available in the market focus on creating isolated components. For creators, there's a significant effort involved in assembling these various creative elements to tell a cohesive story, often facing hurdles with maintaining consistency and facilitating collaboration. CreateIn AI Labs, a trailblazing Creative AI company in Silicon Valley, is revolutionizing the $900 billion content marketing sector. It addresses the challenges of a fragmented market and user confusion and is developing the world's inaugural end to end AI story video creation tool, empowering individuals and businesses to become a compelling storyteller.

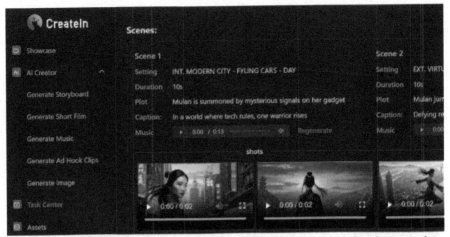

Video storyboard generation tool at CreateIn AI Labs to power creative ideation and collaboration

Work and Productivity

Search & Discovery: Many of us have experienced the familiar challenge of typing a question into Google only to be inundated with an overwhelming barrage of links, often containing conflicting or promotional information. But imagine a different scenario: where you search in whatever channel or medium you want and receive a single, precise answer in plain language, accompanied by optional links for further exploration if desired. Several leaders in the generative AI search space are reshaping how people discover and share information via a chatbot interface. Perplexity AI[46] allows search and query in specific channels (such as YouTube, Academic, Reddit, Wikipedia, etc.) to improve accuracy and relevance of search results. You.com[47] provides personalized in-app search over 150 apps in various categories such as coding, video, education, facts, finance, how-to, news, and places to let you discover more in less time. Consensus[48] improves scientific web search, extracting knowledge from millions of research papers and prioritizing source authority over popularity. Twelve Labs[49] applies AI to make video content searchable, broadening the scope of accessible information.

Let's try an example:

> *Perplexity AI Prompt: Help me find the best hybrid SUV for a family of five with excellent safety ratings and under $40,000*

Perplexity search results curate a list of options based on your specific needs and show you additional resources and related questions to learn more.

Enterprise Search and Knowledge Management: While language models are revolutionizing consumer search, they are also changing enterprise search and knowledge management. Dropbox launched Dropbox Dash[50], an AI-powered universal search that connects all of your tools, content, and apps in a single search bar so you don't need to search for things in multiple places. Glean[51] delivers personalized, permissions-aware enterprise search results; the app knows whether you are a full-time employee, part-time contractor, or admin, and in which departments you work, so you can access certain documents and not others to protect confidential company information. Glean semantically understands natural language queries and adapting to a company's unique database, offering easy setup and scalability with 100+ app connectors, differentiating from its traditional enterprise search competitors (such as Microsoft, Google, Amazon, IBM, Oracle, Coveo, Lucidworks, and Mindbreeze), which tend to focus on single Software-as-a-Service (SaaS) or app search or limited number of SaaS connections.

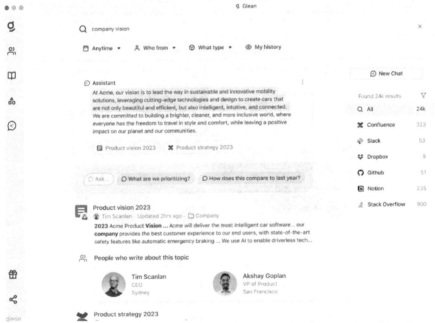

Glean can search across all company's apps to improve productivity and knowledge sharing.

Design: When it comes to PowerPoint storytelling, Tome[52] and Beautiful.ai[53] leverage AI to redefine presentation creation, transforming raw data into compelling narratives and visually appealing slides. In addition, generative AI solutions are dramatically streamlining the design process. Incumbent players like Canva[54] and Microsoft[55] are integrating AI into their existing platforms. Startups like Galileo AI[56] offer an AI copilot for interface design, creating delightful, editable UI designs from a text description. Magician[57] and Uizard[58] enable rapid prototype creation by automating aspects of the design process, allowing designers to focus on strategic aspects of UX design. Monterey[59] leverages AI to enable companies to set up feedback infrastructure, including embeddable widgets, request portals, and connections with customer relationship management (CRM), social media, and more. Monterey therefore helps design a feedback experience that encourages users to share more, which, in turn, helps businesses understand their customers better and make informed strategic decisions.

Uizard's AI Design Assistant can transform your paper sketches into fully functioning UI prototypes.

Sales, Marketing, and Customer Success: Sales and marketing, the lifeblood of every enterprise, is at an exciting crossroads with the advent of generative AI. Companies are no longer relying solely on human insight and intuition, but are increasingly leveraging AI's capabilities to automate, personalize, and optimize their efforts. The customer success landscape is also shifting dramatically, powered by AI-driven technologies that enable personalized, instantaneous, and seamless customer experiences. Sales automation startups like ColdReach.ai[60], Intently.ai[61], Regie.ai[62] and Twain[63] use AI to scale outreach and lead generation, simplifying the process of identifying and engaging potential customers and understanding and engaging customers on a personal level. While Second Nature[64] is using AI to upskill salesforce with AI role-play, Walnut[65] employs AI to create interactive and personalized product demos. Sameday[66] can perform many of the functions of a salesperson, answering phones, responding to emails and chats, and following up with leads.

Startups such as Cresta[67], Ada[68], ASAPP[69], Birch AI[70], Dialpad[71], Forethought[72], Observe.ai[73], and OpenDialog[74] are using AI in myriad ways to enhance customer interactions: these include AI-powered chatbots, intelligence contact centers, and real-time customer support analytics. These startups, and many more, are leading the way in the AI revolution, signaling an exciting future for sales, marketing, and customer success.

Legal, Finance, and Human Resources: The integration of AI in accounting, law, and human resources promises increased efficiency, accuracy, and productivity for business in various sizes. AI software such as Harvey[75], Spellbook[76], and Casetext[77] leverage AI to improve the legal sector. Harvey uses AI to analyze legal contracts and other documents, automating tasks traditionally done by human lawyers and paralegals. Spellbook is designed to democratize legal services by providing an intelligent legal research assistant that can pull relevant case laws, statutes, and regulations; it helps draft legal contracts three times faster by using generative AI to review contracts and suggest appropriate language. Casetext uses AI to simplify legal research; its tools improve search results and can also assist with brief review.

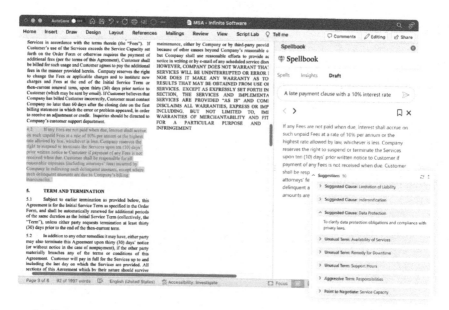

Spellbook's contract writing tool for Microsoft Word facilitates faster and easier contract drafting[76].

In the finance and bookkeeping domains, Truewind[78] and Kick[79] leverage AI to automate and enhance accounting operations. They simplify financial data analysis, provide predictive insights and automate repetitive tasks, allowing financial professionals to focus on strategic decision-making.

When it comes to HR operations and employee communications, several startups are introducing innovative solutions. Effy AI[80] and Onloop[81] utilizes AI to facilitate employee engagement and performance tracking. Paradox[82] is a conversational hiring software company that provides AI-powered tools to streamline and automate HR workflows. Their conversational AI chatbot can screen candidates, schedule interviews, answer questions, collect feedback, and more.

As we usher in a generative AI future, AI's established role in productivity will escalate exponentially. These revolutionary businesses are leading the charge, introducing new ways to automate, streamline, and enhance the ways we work.

Education

Generative AI's impacts can also be felt in the education sector. Imagine a world in which every learner, irrespective of age or location, has a personal language tutor: an AI-powered mentor who converses in real-time, corrects pronunciation, and offers context-specific language usage tips. Poised[83], an AI-powered platform, serves as a personal speech coach, fine-tuning our verbal and non-verbal communication skills to strike the perfect chord in any interaction, be it a professional presentation or a heartfelt conversation.

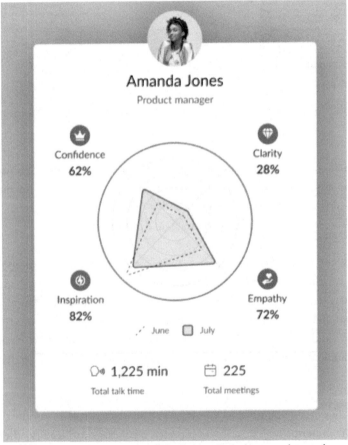

By monitoring how you speak at virtual meetings, Poised can analyze and score how influential and inspiring you are and coach you to speak with more confidence and clarity[83].

The reach of AI doesn't halt at subject-specific learning. It also lends a helping hand to students wrestling with their academic assignments. Grammarly[84] is the virtual equivalent of an encouraging English teacher, helping students tackle writer's block, refine their prose, and elevate their writing style.

The Khan Academy has introduced an experimental AI guide, Khanmigo, to deliver personalized tutoring experiences for learners. Khanmigo[85] offers tailored assistance, encourages critical thinking, and provides pertinent resources while assisting teachers with administrative tasks. Duolingo, the language education specialist, has harnessed the power of GPT-4 to develop two novel features to provide students more feedback and enable improved language practice[86]. Duolingo Max allows students to receive detailed explanations about the correctness of their practice or test answers, mirroring the feedback of a human tutor. The second AI-enabled feature enables language practice through role-playing with AI personas, such as ordering drinks from a barista in a Parisian cafe. The personas each have their own unique personalities and backstories with which learners can interact.

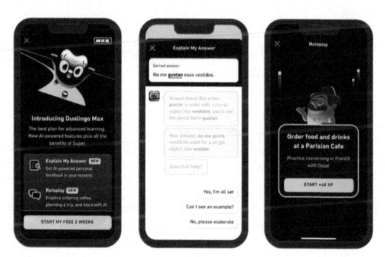

Duolingo Max AI role-playing gamifies immersive and personalized learning. Image from TechCrunch[87].

In higher education, Cheggmate has incorporated GPT-4 into its platform to support college students with their assignments. Similarly, Udacity, an online course provider, has also leveraged GPT-4 to develop an intelligent virtual tutor. This AI-powered tutor provides personalized guidance, offers detailed explanations, simplifies complex concepts, tailoring the learning process to the unique needs of the individual learner. These examples illustrate the expansive role of AI in revolutionizing educational experiences across the spectrum.

Venturing beyond academic learning, we have platforms like Practica[88], which extend the power of AI to the realm of career development. Their custom AI model draws on domain experts and acts like a seasoned mentor, finding the best learning resources and using them to generate personalized next steps to help you get unstuck at work.

As we look to the future, we see a landscape in which generative AI is not merely a tool in education; it is an integral part of the process, a silent partner aiding learners and educators alike[89]. The pedagogical potential of AI transcends geographical boundaries and socioeconomic divisions, illuminating a path towards a world where quality education is not a privilege, but a universally-accessible right.

What Is the Future of Human-to-Human and Human-Machine Relationships in a Generative AI World?

In the melody of life, we often find the sweetest notes in the symphony of human connection. As the generative AI era unfolds, we stand on the precipice of a new harmony: one in which our uniquely human chords resonate seamlessly with the powerful rhythm of artificial intelligence.

In the interplay between human and artificial intelligence, boundaries between work and life blur. AI has progressively become a supportive player in enhancing emotional wellbeing. Leading the charge in this sentiment-focused progression is Inflection AI[90], an AI-powered platform designed to meet our emotional needs. Inflection AI's chatbot, Pi, provides a comforting presence, offering therapeutic exchanges during solitary nights and challenging days. It's a listening, understanding, and comforting companion always available at your fingertips. AI companions like Replika[91] and Character.ai[92] have also established their niche.

They operate as AI allies, providing safe spaces for users to express their most profound fears, loftiest aspirations, and everything that falls in between. They reflect the intricate dynamics that sustain our most valued human relationships.

Meanwhile, platforms like Woebot[93] and Wysa[94] have achieved clinically-proven outcomes and have earned FDA approvals for their effectiveness in managing some mental health issues. They act as wellness coaches, providing personalized, therapeutic interaction and mental health assistance, constantly reminding us that we're never truly alone. Meeno, led by former Tinder CEO Renate Nyborg and backed by Andrew Ng's AI Fund, seeks to leverage the power of AI to help one billion people, and especially young adults, master the art of human connection[95].

Looking forward, we envision generative AI not as a mechanistic entity, but as a luminous force promoting improved human connection. It breathes new life into our interactions, deepening our emotional connections and enriching the core of our personal and professional exchanges.

In the era of generative AI, we're not mere passive observers witnessing AI redefining our relationships. Instead, we are active participants, shaping the rhythm of our connections with every interaction. We're at a juncture where AI's potency doesn't supplant the core of human connection but enhances and amplifies it. The result is a harmonious, symphonic blend of technology and humanity, reverberating with the potential of a more connected future.

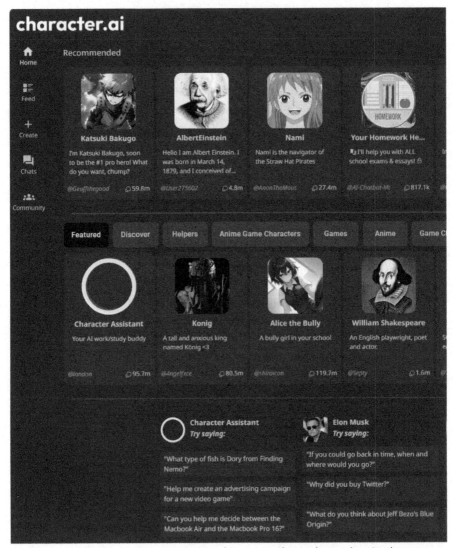

Character AI allows you to create a new character or chat with your favorite characters,
fictional or real, dead or alive[92].

A screenshot from the Replika app. Your Replika will always be by your side. Chat about your day, do fun or relaxing activities together, share real-life experiences in augmented reality, catch up on video calls, and more.

1.42 - What's Next for the Shared Future of Generative AI and Robotics?

A rapidly growing field of study is embodied AI: developing intelligent agents that can interact with the physical world. These agents are designed to have a visual representation within an environment, be it real or virtual, and are optimizing to perform the role of a persona (e.g., bank teller, travel assistant, etc.). Embodied AI combines computer vision, language, graphics, and robotics to create agents that can perform a wide range of tasks, from simple object recognition to complex decision-making processes.

Recent advances in deep learning, reinforcement learning, and generative AI have enabled researchers to make rapid progress towards creating intelligent agents that can interact with the physical world in a meaningful way. In this section, we will explore the forefront of innovation in embodied AI, focusing on generative AI agents and robotics.

Generative AI Agents

An AI agent, often termed as an AI "copilot," is an intelligent system capable of performing complex tasks, requiring high-level reasoning, memory, planning, and the ability to interact with the world in a meaningful way.

The work of researchers like Lilian Weng from OpenAI has shed light on the structure and components of a Large Language Model (LLM)-powered agent system. Within this system, LLM functions like the agent's brain, working in harmony with key components such as planning structures, memory mechanisms, and tools.

Planning structures enable the agent to break down complex tasks into smaller subgoals and to reflect on past actions for improved future performance. Memory mechanisms, both short and long-term, allow the agent to learn from prompts and retain information over extended periods. Finally, tools allow the AI agent to access external information and resources, enhancing its problem-solving capabilities to take necessary executable actions.

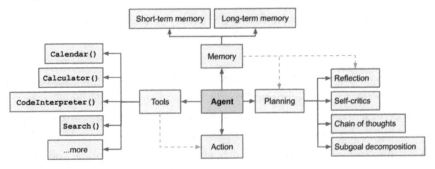

Overview of a LLM-powered autonomous agent system by Lilian Weng from Open AI. Image from Lilian Weng[96].

Given that the present iteration of ChatGPT still relies on manual human input for each task, there is a rising interest in Autonomous Generative Agents capable of managing end-to-end workflow. The study *"Generative Agents"* by researchers at Stanford University and tech giant Google (Park et al., 2023) unveils a fascinating experiment in which 25 virtual characters, each guided by an LLM-powered agent, interact within a sandbox setting, mimicking human behaviors with capabilities of memory, planning, and reflection.

AI agents can perform a variety of tasks, ranging from simpler tasks like ordering food online to more complex ones such as creating investment strategies or summarizing meetings. The excitement in the AI space is palpable with the advent of proof-of-concept projects such as AutoGPT, BabyAGI, and CAMEL (Communicative Agents for "Mind" Exploration of Large Scale Language Model Society). However, as much as these early demonstrations are inspiring, developers also acknowledge the inherent risks and challenges.

To deepen our understanding of "Generative AI Agents," let's examine a diagram by Kenn So, an investor and writer at the blog Generational. This visualization dissects the elements of cognitive architecture in the human mind and devises ways that AI can be used to create an artificial mind — the foremost big-picture goal in the AI world.

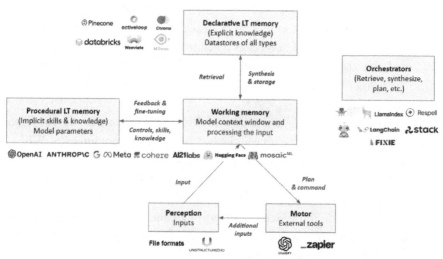

Schematic of a cognitive model of computing from Kenn So's newsletter[97].

Kenn So proposed a general cognitive model with six main components: perception, short-term working memory, two kinds of long-term memory (procedural and declarative memory), motor function, and an orchestrator function that manages it all:

- **Perception**: Perception is the AI's ability to understand and process sensory data, similar to human senses. Think of this as AI's front-end data collection and initial processing layer—e.g., NLP algorithms for chatbots or computer vision for image recognition.

- **Working Memory**: This is a temporary holding area where transient data gets processed. In the context of generative AI, it can be likened to the context window in LLMs, storing immediate interactions for quick referencing.

- **Procedural Long-term Memory**: This component is akin to a repository for the AI's skills and habits. These are the built-in rules or routines that your product follows, such as ethical guidelines and automated responses.

- **Declarative Long-term Memory:** This is the database where facts and events are stored. Product teams might use different kinds of databases to facilitate this, like knowledge graphs for storing interconnected facts or vector databases for numerical data.

- **Motor:** Motor functions in AI refer to its ability to interact with external systems, like sending an email or making an API call.

- **Orchestrator**: This component coordinates all the other elements. This could be the system that dictates when to pull data from databases, how to feed the data into the working memory, or how to implement task management.

When it comes to thinking about how it all works together, the Orchestrator acts as a conductor, directing how and when each component interacts. When a user engages with your generative AI product, Perception kicks in to understand the user input. Working Memory temporarily stores this data while the Orchestrator pulls relevant information from Declarative Long-term Memory. The Procedural Long-term Memory offers guidelines on how to interact, while the Motor executes any required actions. Thus, all the elements are continuously in sync, providing a cohesive user experience[98].

Examples of AI Agents

The journey towards AI agents has been significantly accelerated with the release of OpenAI's GPT-4 model. Its capacity to facilitate strategic and adaptable thinking opens the possibility of navigating real-world unpredictability. The Microsoft 365 Copilot[99] aims to transform work in three ways: unleash creativity, unlock productivity, and uplevel skills. With Copilot in Microsoft Word, users can jump-start the creative writing process with a first draft for further editing, so they never have to start with a blank slate again. Copilot in PowerPoint helps create beautiful presentations with a simple prompt, referencing relevant content from past documents. Copilot in Excel can help analyze trends and create data visuals in seconds. In Outlook, Copilot lightens the workload in myriads of ways: from summarizing long email threads to drafting suggested replies. Its tasks in Teams range from capturing key discussion points in live meetings to suggesting action items. When it comes to Viva, Copilot does everything from generating personalized learning journeys to discovering relevant resources and scheduling trainings. GitHub's Copilot feature[100] uses machine learning algorithms to recommend code changes and optimizations, providing developers with valuable insights to improve their work. According to Microsoft, among developers using GitHub Copilot, 88% have reported increased productivity, 74% claim it enables them to concentrate on more fulfilling tasks, and 77% believe it reduces the time they spend searching for information or examples. Beyond just boosting individual efficiency, Copilot introduces a novel knowledge model for organizations, effectively utilizing the vast pool of data and insights that are currently underutilized and hard to access inside a company.

Additional examples of AI agents from tech incumbents include Zoom's Transcription Service[101] that uses AI to transcribe meeting recordings, enhancing note-taking and accessibility. Salesforce's Einstein GPT[102] provides AI-driven insights and recommendations to sales and marketing teams. Adobe Sensei[103] enhances creative workflows and content creation tasks in Adobe's suite of creative software. Asana Intelligence[104] uses AI to intelligently assign tasks to team members based on workload and expertise and serve as a writing assistant. In the startup AI landscape, there's a surge of AI team assistant tools aimed at automating and optimizing note-taking and task management. Adept[105] is aiming to build an AI team assistant designed to boost team efficiency and productivity. Cogram[106] and Sembly[107] serve as your digital team assistant to take notes in virtual meetings, track action items, and automate downstream tasks, while keeping your data private and secure. Meanwhile, the Gist[108] provides summaries of Slack discussions.

Limitations of AI Agents

Nonetheless, developers recognize that today's agents are just proof-of-concepts and can often propose illogical solutions. Indeed, the industry is aware of the ethical challenges associated with AI development. Key concerns include the fears of replicating human biases, potential for misinformation, and potential for harm, either accidental or deliberate. We dedicated section 2.25 to talk about how to build Responsible AI applications and other legal, ethical, and social implications to watch out for.

In this landscape, many generative AI applications will revolutionize the way people and AI interact, fundamentally shifting our work patterns. Like any transformative work practice, there's an inherent learning curve. However, those who adapt and embrace this emerging way of working will swiftly acquire a competitive advantage. The future of work lies not in AI replacing humans but in a synergistic collaboration that amplifies our potential. While this technology isn't entirely mature yet, its evolution is certainly steering us towards an exciting new era of amplified human potential.

Generative AI and Robotics

Stanford University and University of Illinois Urbana-Champaign recently unveiled groundbreaking research paper ("VoxPoser: Composable 3D Value Maps for Robotic Manipulation with Language Models" authored by Wenlong Huang, Chen Wang, Ruohan Zhang, Yunzhu Li, Jiajun Wu, and Li Fei-Fei) in embodied intelligence, in which LLMs were integrated with robots. This innovative approach allowed complex instructions to be converted into specific action plans without the need for additional data or training.

This development in AI and robotics enables humans to issue instructions to robots using natural language. For instance, a user could direct a robot to "Open the drawer above, carefully avoiding the vase!" The integration of large language models and visual language models enables robots to analyze objectives and obstacles in 3D space, assisting in action planning.

The exciting part of this research is that it allows robots to perform tasks in the real world without any prior "training" – what's known as zero-shot synthesis. In other words, the researchers have come up with a method that enables a robot to do something it has never done before, without needing any demonstration beforehand. They called this system VoxPoser[109].

The idea is simple. When the robot is given information about its surroundings – gathered through RGB-D imaging – and instructions in plain language, the LLM generates a special type of code based on this information. This code then works with the visual language model (VLM) to create a set of operation instructions. This set of instructions tells the robot where and how to act. Compared to older methods which needed additional training, this new approach addresses the lack of training data for robots.

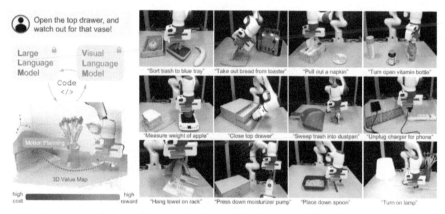

VoxPoser can take instructions given in everyday language and use them to decide what actions a robot can and cannot perform. It does this by connecting these instructions to what the robot can see around it, without needing any extra training. The end result is a kind of 3D map that helps the robot know what to do. This way, the robot can carry out a wide range of tasks it's never done before, using instructions it hasn't heard before and interacting with objects it hasn't encountered before.

This study not only showcases a path forward in the realm of AI and robotics but also illustrates how large models could help solve complex problems and navigate real-world tasks. 1X Robotics[110], previously known as Halodi Robotics, is a Norwegian company specializing in the creation of humanoid robots capable of human-like movement and actions. They recently received a substantial investment of $23.5 million from OpenAI's startup fund. Tesla, under the leadership of Elon Musk, also revealed their humanoid robot named the Tesla Bot or Optimus in 2021[111]. As we look ahead, the collaboration between AI and robotics continues to herald a new era of technological advancement and possibilities.

1.5 - Limitations of Present-Day Generative AI

1.51 - What Can Today's Generative AI Technology Truly Achieve?

It's tempting to view AI as a panacea for many of our problems. However, a clear-eyed assessment helps us discern the areas where AI shines from those where it still falters:

What Can Generative AI Do vs. Not Do

Aspect	What Generative AI Can Do	What Generative AI Can Not Yet Do
Language & Communication	Understand and generate human-like text based on given prompts. Translate languages.	Fully grasp nuances, emotions, and deep cultural contexts in communication.
Data Processing	Analyze vast datasets and identify patterns efficiently.	Intuitively identify incorrect or biased data. Understand the real-world implications of data trends.
Learning & Adaptability	Learn and improve from new data through training. Adapt to new tasks with transfer learning (to a degree).	Learn and adapt from limited data or experiences as humans do. Generalize learning across vastly different domains.
Creativity & Innovation	Generate creative outputs in art, music, etc. based on existing patterns and data.	Truly innovate or think outside the box without a data-driven precedent.
Task Execution	Perform specific tasks consistently and efficiently. Automate routine and predefined tasks.	Understand the broader context or purpose of tasks beyond its programmed scope. Perform unexpected tasks or tasks requiring nuanced human judgment or physical interaction.
Emotion & Empathy	Detect patterns that may indicate emotions (e.g., sentiment analysis).	Truly feel emotions or exhibit genuine empathy. Adapt emotional responses to nuanced social contexts.
Decision-Making	Make decisions based on data-driven logic. Optimize decisions based on statistical analysis	Always account for the myriad of human factors (ethical, emotional) in decisions. Understand the broader implications of decisions.
Generalization	Operate effectively within its trained domain. Apply learned patterns to similar domains (to a degree).	Broadly transfer knowledge across vastly different domains without additional training.
Ethics & Values	Follow guidelines and rules set by programmers.	Innately understand and apply human values, ethics, or morals.

Copyright ©2024 by Reimagined Authors Shi, Cai, and Rong

In essence, while AI offers unparalleled advantages in data processing, consistency, and task-specific operations, it may lack the deep contextual understanding and emotional intelligence that are second nature to humans. The challenge, therefore, lies in harnessing AI's strengths while being acutely aware of its limitations.

1.52 - What Are the Areas to Watch Out for When Working with Generative AI?

In the rapidly evolving world of generative AI, it's essential to be mindful of its challenges and limitations, balancing the excitement of potential breakthroughs with a grounded understanding of its current state.

The task of creating a coherent sentence, seemingly simple for humans, exemplifies the complexity generative AI faces in mastering language intricacies. Despite advancements, AI models still struggle with aspects like **lexicon, grammar, context, and tonality**.

A paramount challenge lies in **infusing a desired degree of control and determinism** into the generative process. Unlike humans, these AI models lack an innate grasp of human values, making it difficult to tailor their outputs to our subjective preferences, such as humor or ethics. Balancing creative freedom with appropriate control is a nuanced challenge for engineers.

Maintaining consistent output quality is another issue. For instance, GANs might produce visually striking images marred by imperfections. Language models can generate text that lacks clarity or consistency upon detailed examination, revealing a gap in quality assurance.

In addition, current AI approaches remain data-hungry, **requiring vast amounts of high-quality data for training**. Despite the advent of few-shot and transfer learning, the procurement and curation of expansive datasets are still resource-draining. Furthermore, there is the added challenge of ensuring that this **data is representative and diverse**, to prevent unintentional biases from infiltrating AI models given the nuances of human experiences.

Digging deeper, the **lack of memory** in generative AI technology, or the **context limit**, restricts the AI's grasp of preceding interactions, often leading to a disjointed and incoherent dialogue flow. This is a stark contrast to human communicators who effortlessly reference past exchanges to enrich conversations. This challenge is being actively addressed.

A looming concern is the potential **erosion of content originality and diversity of perspectives**, as generative AI tends to regurgitate a homogenized blend of ideas based on its training data. The cycle of AI training on AI-generated content risks a "cookie-cutter" effect, potentially stifling fresh, divergent thoughts. This, paired with an over-reliance on AI for ideation, might lead to a tangible loss of creativity, as the human impulse to innovate gets overshadowed by AI's prompt suggestions.

Moreover, the ease of generating plausible yet unauthentic content paves the way for **academic dishonesty and plagiarism**, as students might lean on AI to complete assignments, thereby stunting the growth of their critical thinking muscles.

On a broader societal canvas, the specter of **job displacement** looms large as generative AI automates a widening array of tasks. Coupled with the existential anxiety stemming from a future where **AI-driven decision-making eclipses human judgment**, we are nudged to ponder on the evolving essence of our purpose.

Adding to these concerns, a New York Times article published in December 2023 featured a troubling AI concept, **P(doom)** - a metric indicating the "probability of doom" associated with AI[188]. This metric reflects AI researchers' assessments of the likelihood that AI could lead to catastrophic outcomes for humanity. It's crucial to acknowledge, however, that there is no universal consensus on the actual P(doom) value, as opinions vary widely among experts. Notably, Geoffrey Hinton, a renowned AI researcher who left Google in 2023, estimated a 10% chance of human extinction due to unregulated AI in the next 30 years[188]. Echoing these concerns, Elon Musk stated, "In my view, AI is more dangerous than nuclear bombs, and we regulate nuclear bombs. You can't just go make a nuclear bomb in your backyard. I think we should have some kind of regulation with AI.[189]" These statements underscore the urgent need for careful consideration and regulation in the field of AI to prevent potentially dire consequences.

Current Challenges, Limitations & Implications of Generative AI

Current Challenges, Limitations, and Implications of Generative AI

Dimension	Considerations
Data Quality	☐ *Data Dependency*: Generative AI requires vast amounts of quality data ☐ *Garbage In, Garbage Out*: Poor quality data leads to flawed outputs ☐ *Overfitting and Underfitting*: Inaccurate models due to poor quality data
Privacy	☐ *Data Collection*: Large scale data collection may infringe privacy ☐ *Anonymity*: Risk of generating data traceable to real individuals
Legal	☐ *Copyright Infringement*: Potential infringement through AI-generated content ☐ *Data Regulations*: Compliance with data laws like General Data Protection Regulation (GDPR) and California Consumer Privacy Act (CCPA)
Social & Emotional	☐ *Job Displacement*: Automation risks resulting in job loss ☐ *Misinformation*: Risk of spreading false information or 'deepfakes' ☐ *Lack of Creativity and Cognitive Diversity*: Limitation to generate only based on seen patterns, online echo chambers reinforcing existing beliefs and preferences
Ethical	☐ *Bias and Discrimination*: Propagation of biases (such as gender, race, or socio-economic biases) leading to unfair or discriminatory outcomes. ☐ *Consent*: Ensuring users' informed consent ☐ *Transparency and Explainability*: Understanding the decision-making process of AI models – in other words, looking into the "black box"
Technical Challenges	☐ *Computational Resources*: High demand for processing power and memory ☐ *Model Transparency*: Complex and opaque decision-making process ☐ *Evaluation Difficulty*: Quantitatively assessing the quality of AI outputs ☐ *Lack of Control*: Difficulty in controlling the outputs of generative models ☐ *Error Propagation*: Accumulation of errors in sequence generation tasks ☐ *Resource Inefficiency*: Inefficient use of resources for marginal improvement in output quality

Copyright ©2024 by Reimagined Authors Shi, Cai, and Rong

Considering these challenges and limitations, we must face the inevitable truth that generative AI is far from perfect. Yet, as we immerse ourselves in this brave new world of synthetic media and hyper-realistic simulations, it is important to remember that these hurdles are not insurmountable. Guided by the spirit of innovation that lies at the heart of human progress, we will continue to refine the scaffolding of generative AI techniques, with each iteration bringing us closer to the visage of machines possessing the ability to evoke awe, wonder, and delight.

Part II: Building Generative AI Products

As we transition from understanding the vast landscape of generative AI to building delightful and responsible products in Part II, we will begin the journey with a better understanding of our customers and their needs.

2.1 - Whose Problem Are We Really Solving?

2.11 - Why Do AI Products Often Miss the Mark on Customer Segmentation?

Every week, our email inboxes are flooded with pitches that go something like this: "Dear [Name], we're an up-and-coming AI startup leveraging cutting-edge Llama/Stable Diffusion algorithms, fine-tuned on proprietary datasets that no one has access to. We're on the cusp of a breakthrough, and we believe everyone will be our customer. Would you like to be our inaugural product manager?"

Would an experienced product manager jump at this offer? Many experienced product managers certainly wouldn't.

So, what's wrong here?

The fundamental mistake here lies in the lack of segmentation. When business leaders set out to create a product, their first imperative should be to identify the target customer base they aim to serve. The notion of "everyone as a potential customer" isn't just naïve—it's a glaring signal of an absence of strategic product leadership. No product can be all things to all people; the key is to pinpoint the specific market segment where you can provide unique value.

What Is the Right Way to Segment Your Customers?

Segmenting your customer base is a critical step in tailoring your product to meet specific needs. Here are some proven approaches to customer segmentation:

1. **Needs-Based Segmentation**: This is often the starting point, focusing on particular problems or needs your product addresses. We will elaborate more about this approach in the subsequent section where we will also introduce the Jobs-to-Be-Done (JTBD) framework.
   - *Example*: Generative AI can alleviate writer's block for content creators, automate initial designs for designers, or streamline code generation for developers. Keep in mind that AI output may require further refinement.

2. **Demographic Segmentation**: This complements a needs-based approach, targeting specific age groups, genders, income levels, educational backgrounds, and occupations.
   - *Example*: A generative AI for personalized fitness plans might target affluent, middle-aged individuals. AI should be designed to avoid gender bias.

3. **Job Role Segmentation (B2B)**: In a business setting, consider the specific roles and responsibilities within target companies.
   - *Example*: Researchers may use generative AI for data analysis and report generation, while marketers could use it to produce personalized ad content and run A/B tests.

4. **Industry and Company-Attribute Segmentation (B2B)**: Further refine your B2B targeting by considering the industry, company size, and stage of growth.
   - *Example*: A generative AI for legal documents should be well-versed in legal jargon and comply with regulations. Consider integrating the AI into existing systems for better adoption in law firms and corporate legal departments (versus as a standalone application).

5. **Psychographic Segmentation**: This involves categorizing customers based on personal traits like values, interests, and technology adoption readiness.
   o *Example*: For an early-stage generative AI product, the target demographic could be tech enthusiasts who are open to new technologies and likely to provide valuable feedback. For mass-market applications, focus on user-friendly design and robust customer support.

By employing these segmentation strategies, you'll be better positioned to create a product that resonates with your target market, thus increasing your chances of success.

How to Choose the Right Segment to Focus?

Navigating the maze of customer segmentation is more complex than textbook knowledge or coaching advice. Here's a playbook to guide your decision-making in choosing the right customer segment for your generative AI application:

Assessing Segment Viability: Balancing Attractiveness and AI Trust

Once you've segmented your audience, you'll need to weigh various factors to identify your prime target. These include market size, resonance with value proposition, willingness to pay, alignment with company vision, ability to reach the target customer, and their trust and readiness to adopt AI.

Market Size

The potential market size is a balancing act. A larger segment promises a wider customer base but may be more competitive or diluted in need. A niche market, on the other hand, may offer easier market penetration and leadership opportunities. For generative AI, the prevalence of the specific issues your tool solves can also influence market size.

Resonance with Value Proposition

Your ideal customer segment should find your value proposition compelling. If your AI tool excels in generating marketing content, for instance, it will be far more appealing to marketers facing content scalability issues than to other groups.

Willingness to Pay

Your target customers should see enough value in your product to warrant its cost. Assessing this willingness often involves market research, consumer surveys, and behavioral analysis in comparable markets.

Alignment with Company Vision

The chosen segment should resonate with your company's mission and vision. If your company aims to empower content creators, for instance, focus on those who produce content at scale rather than casual users.

Ability to Reach Target Customer

Evaluate how effectively and efficiently it is to reach your target segment. Do you have the necessary marketing channels or partnerships? For B2B products aimed at large corporations, assess your sales capabilities to navigate complex procurement processes.

AI Trust and Adoption Readiness

Generative AI is still a burgeoning field. Trust varies widely across segments, influenced by factors like prior AI experiences, media portrayal, and the user's AI literacy. Building trust requires transparency about your AI's functionality, performance reliability, robust security measures, and exceptional customer support.

In summary, the success of your generative AI application hinges not just on the application's technical prowess but also on its fit with the right customer segment. Use this playbook as your foundation, but remain agile, adapting your strategies as you gather more insights.

Case in Point: How Synthesia Nailed Segment Selection

Picture this: Three computer vision PhD students at the University of Edinburgh – Victor Riparbelli, Marc Skarbek, and Gregor McEwan – are engrossed in an animated discussion. The subject? Generative adversarial networks (GANs) and the untapped potential of digital humans. As the conversation grows more passionate, a light bulb goes off. They realize they're sitting on a potential goldmine: a technology that could revolutionize the way we consume media.

Fast forward to 2017. The trio launches Synthesia, diving headfirst into the development of algorithms that can generate hyper-realistic videos. Their groundbreaking innovation? AI agents—digital doubles that could mimic real people down to their voice and mannerisms. Imagine a digital David Beckham who could say anything you programmed him to say, in his own recognizable voice.

When 2019 rolls around, Synthesia steps out of the shadows, unveiling demos that feature AI avatars of iconic figures like Beckham. The tech world takes notice, and investors are quick to jump on this high-potential venture.

Here's the kicker, though: Synthesia's success lies not just in the caliber of their technology, which stands on its own merits, but also in their strategic focus on a niche market—educational and marketing videos requiring template-based solutions. They have expertly identified and served a customer segment with a distinct, pressing need, perfectly matching their robust technical capabilities to address it.

In essence, Synthesia's story isn't just about groundbreaking technology; it's a textbook example of the power of choosing the right customer segment. They identified a niche where the demand was high, the pain was acute, and a quality solution was not just desired but desperately needed – and they delivered, catapulting themselves to the forefront of AI video generation.

So, when you think about how to position your generative AI application, remember the Synthesia story. It's not always about having the most advanced technology; sometimes, it's about knowing exactly who needs it the most.

2.12 - Problem First or Tech First? The Dilemma in AI Problem Identification

Innovating with Empathy: Apple's Approach to Tech Revolution
Consider the technology giants like Apple and their most iconic products: the Mac with its Graphic User Interface (GUI), the iPod, and the iPhone. A surprising fact is that Apple wasn't the original inventor of these technologies. The GUI originated from Xerox, while the concept of the MP3 player was first developed by Korea's Saehan Information Systems. Despite this, Apple succeeded in transforming these technologies into the cornerstone of a multi-trillion-dollar empire.

Apple excels in understanding the core problems consumers face and then passionately crafting solutions that speak directly to those needs. In the realm of technological innovation, while individual technologies may ebb and flow, the power lies in a company's capacity to deeply understand and creatively solve the problems these technologies address.

This principle holds true for AI as well. Generative AI has the potential to revolutionize industries across the board. Yet, the fatal mistake is to build technology first and then go hunting for a problem it can solve. True success comes when you flip the script: start with the problem that begs to be solved and let that guide your technological innovation.

How do we identify the problem effectively? One useful framework is the Jobs-to-Be-Done (JTBD).

Uncovering Jobs-to-Be-Done (JTBD): Rooting AI in Real User Needs

The Jobs to Be Done (JTBD) framework is a powerful tool that product managers can employ to gain deep insights into the needs and motivations of users. It focuses on the core tasks that your customers are trying to accomplish. Essentially, what is the "job" that your product is being "hired" to do? Consider, for example, that people who need to drill a hole often do not want a drill; they just want the hole to be in the wall. Another example: a user might hire a video-streaming app like Netflix or Hulu, not just to watch shows, but really to entertain themselves and unwind after a long day. Understanding this can help in assessing if AI can further enhance the solution to the job or if it can innovate a new way to get the job done.

A famous case study that perfectly illustrates the JTBD framework is the milkshake study. McDonald's hired Harvard Professor Clayton M. Christensen and his team to conduct research to boost milkshake sales. McDonald's first took a classic approach and did the obvious: they sought to increase sales by surveying their milkshake-buying customers, and incorporating their feedback to make the milkshakes cooler, more chocolatey, and sweeter. Despite this, sales remained flat.

Instead of focusing on flavor preferences, Christensen and his colleagues studied why the customers were buying milkshakes at a particular time of the day. They noticed that 40% of sales were made in the morning by lone males who ordered nothing but milkshakes to consume in their cars. Those customers bought milkshakes to occupy themselves during a long, boring commute every morning. They chose milkshakes over bagels or bananas, because milkshakes aren't messy, could be wrangled with one hand, and took a considerable time to ingest: perfect to combat that 10 a.m. mid-morning hunger attack.

What was the "Job to Be Done" or JTBD here? In this case, it was to satisfy hunger during the long morning commute.

Armed with this insight, McDonald's made their milkshakes even thicker, swirling in tiny chunks of fruit to introduce unpredictability, and moving dispensing machines in front of the counter to help customers avoid drive-thru lines. Sales skyrocketed.

Case in Point: How Intercom Transformed Its Go-to-Market Through Jobs-to-Be-Done

Imagine the company Intercom at a crossroads. After rapidly climbing the SaaS ladder to hit a $100M annual recurring revenue (ARR), they found themselves pondering their next strategic move. Traditional persona-based marketing had taken them far, but the team sensed an impending plateau. The eureka moment came when they flipped the script and asked, "What job is our product being hired to do?"

Instead of selling a Swiss Army knife, they decided to offer the perfect tool for each specific job. They segmented their single product into four specialized solutions, each finely tuned to solve a particular customer's job.'

To dig deep into these 'jobs,' Intercom didn't just rely on data and analytics; they went straight to the source. They conversed with 40 different customers, from excited newcomers to those who had recently clicked 'cancel.' These candid dialogues unearthed compelling patterns and crystallized into actionable JTBD insights.

The company's homepage in 2015 aptly showcased this shift. Gone were the generic pitches; they were replaced by five unique offerings, each spotlighted for the specific 'job' it set out to accomplish. This wasn't just a marketing gimmick—it was a strategic realignment. It influenced not just how they talked about their products, but how they developed them. It was more than a change; it was a revolution. And it all started with asking the right question: "What job are you hiring Intercom to do?"

Intercom's annual revenue skyrocketed, exceeding 150% during 2015 and 2016, following the adoption of job-based marketing strategies. They increased their revenue by 15x over the subsequent eight-year period.

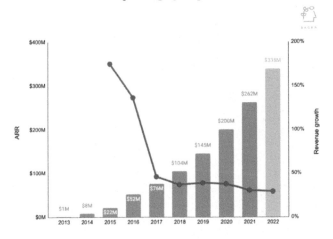

Annual Recurring Revenue over time for Intercom. Image from UX Planet[112].

Understanding Circumstances to Deepen Empathy

The McDonalds milkshake study and the Intercom example serve as reminders that true innovation blossoms when we understand the core problem or job the customer aims to solve. In the realm of generative AI, these insights are doubly crucial. The technology is emerging, and user expectations are still being formed.

Features
What People Buy

Jobs To Be Done
What People Really Want

Copyright ©2024 by Authors of Reimagined.

Here's how you can break down the types of 'jobs' your generative AI product may serve:

Functional Jobs

Functional jobs are the practical, goal-oriented tasks users aim to complete. These can range from finding quick information to enhancing productivity. In the context of generative AI, functional jobs often revolve around automating tasks, generating insights, or making processes more efficient.

Unique Considerations for Generative AI:

- **Safety and Trust**: If the functional job involves handling sensitive information, your AI must be compliant with privacy laws. You also need to earn the customer's trust by properly handling the data.
- **Quality of Output**: The AI should generate high-quality content, code, or insights that require minimal post-generation editing.

- **Adaptability and Versatility**: The AI should be flexible enough to handle a range of tasks within the functional job for which it's designed, whether it's drafting emails or creating marketing copy.
- **Speed and Efficiency**: The AI should perform tasks faster than a human could, without sacrificing quality.

Social Jobs

Social jobs concern how the product enables users to fit in or distinguish themselves within their social or professional circles. This might involve forming connections, building a sense of community, or receiving support from others.

Unique Considerations for Generative AI:

- **Collaboration**: If the AI tool is used in team settings, it should facilitate rather than hinder collaboration.
- **Status Signaling**: For some users, using a cutting-edge AI tool may serve as a status symbol in their professional community.
- **Community Building**: Are there features, like generative conversation starters or shared templates, that can create community among users?

Emotional Jobs

Emotional jobs relate to the user's emotional state when using your product, be it seeking confidence, control, companionship / social approval, or entertainment.

Unique Considerations for Generative AI:

- **Trustworthiness**: Given the newness and complexity of AI, building trust is crucial. Transparently communicate how the AI makes decisions.
- **Empowerment**: The AI should make users feel empowered, not replaced. Features like customization can give users a sense of control.
- **Reducing Anxiety**: Given that AI can handle complex tasks, consider this question to improve your user's overall experience: how does your product reduce stress or anxiety linked to those tasks?

Identifying functional, social, and emotional jobs is a cornerstone in the development of any product, especially one as nuanced as generative AI. These jobs offer a window into your users' objectives, social ambitions, and emotional triggers. As we dive deeper, we'll explore how to turn these insights into actionable steps for creating user personas and opportunity statements.

The Opportunity Statement: Defining the 'Who' and 'What'

An opportunity statement is a powerful tool that succinctly expresses the value that a product aims to deliver. It helps in keeping the product development focused and customer-centric. When building generative AI products, an opportunity statement can help in highlighting how AI can generate value for users. By integrating the JTBD framework, the Opportunity Statement becomes even more aligned with user needs. Here's how you can craft an Opportunity Statement with JTBD for generative AI products.

The Opportunity Statement Formula

As an [audience + circumstance], I want to [priority JTBD: can have functional, social, or emotional circumstances] in a way that [success criteria].

- **Audience + circumstance**: This is the target audience, along with the specific situations or circumstances that create the need for the product.
- **Priority JTBD**: The core job that the user wants to get done, which can be functional, social, or emotional.
- **Success criteria**: Outline what the user considers to be a successful outcome.

Let's apply the Opportunity Statement Formula to three examples:

Example 1: Perplexity AI (Management Consultant)
Opportunity Statement:
- *[audience + circumstance]* As a management consultant working with varied industries and in need of accurate data-driven insights with valid sources,
- *[priority JTBD]* I want an AI tool that amplifies my research capabilities, refines the quality of my industry assessments, and offers feedback.

- *[success criteria]* My success criterion is having a seamless integration into my consulting process and ensuring the derived insights are both innovative and actionable, significantly reducing research time with improved quality.

Example 2: Khan Academy AI Tutor (Students)
Opportunity Statement:
- *[audience + circumstance]* As a student struggling with specific academic subjects and seeking personalized guidance,
- *[priority JTBD]* I desire an AI tutor that adapts to my learning pace, identifies my weak areas, and offers tailored lessons.
- *[success criteria]* Success for me is reflected in improved comprehension, better grades, and increased confidence in tackling challenging topics.

Example 3: CaseText (Lawyer)
Opportunity Statement:
- *[audience + circumstance]* As a lawyer handling multiple cases simultaneously and in need of quick legal precedents,
- *[priority JTBD]* I seek an AI tool that streamlines my research process, offers precise case recommendations, and provides insights into legal strategies.
- *[success criteria]* For me, success means saving time on research, ensuring case accuracy, and strengthening my case arguments.

Crafting an opportunity statement with JTBD is an essential step in ensuring that your generative AI product is aligned with the needs and desires of your target audience. It serves as a guiding star, helping to focus the product development process on delivering real value to customers. By articulating the users' JTBDs and defining success criteria, product managers can make informed decisions that steer the development of generative AI products towards success.

The Contrarian View: When Prioritizing Tech Can Make Sense

Common wisdom in product management advises a customer-first approach: understand the problem before developing the solution. This ensures that the product serves an actual need rather than becoming a solution seeking a problem. There's an alternate viewpoint that gains prominence particularly in the realm of AI product development. When technology experiences a seismic shift, as is the case with advancements in generative AI, focusing on the technology itself can offer unparalleled long-term ROI. In such scenarios, the cutting-edge tech can expose issues customers didn't even know they had, thereby creating entirely new markets or applications. For example, an AI initially designed for predicting equipment failure in factories could unexpectedly also optimize energy use and production efficiency, revealing needs that companies didn't even know they had. In addition, in many B2B use cases, the AI needs to be mature enough to render real commercial value for enterprises.

Checklist for Making Big Upfront Tech Investments

Below are 5 key factors to help decide whether you should invest more in technology while keeping your ears tuned to customer needs:

1. **Breakthrough Potential:** Is your technology a game-changer? If you're bringing something revolutionary to the table, being first can secure a competitive edge.

2. **High Barriers, Big Payoffs:** If your tech is hard to copy and has the potential to dominate a niche market, the cost of a tech-first approach may be well-justified.

3. **Resource Ready:** Ample capital and top-tier talent give you the runway to focus on R&D without immediate pressure to find a market fit.

4. **Built to Scale:** Is your core tech modular and can it pivot to new applications? If so, upfront investment can provide long-term advantages.

5. **Fast Movers Win:** In fast-evolving sectors, waiting can mean obsolescence. If you're in such a space, speed and agility are key; invest heavily upfront to keep pace.

Use this simple checklist as a strategic tool when debating between a problem-first and tech-first approach for your generative AI product. It's not about throwing customer problems to the wind; rather, it's about knowing when the tech itself is the wind that will carry your ship forward.

In summary, the 'Problem-First' approach continues to be a robust starting point for most product management endeavors, including those in AI. However, when it comes to generative AI products, don't shy away from 'Tech-First' when circumstances warrant it. Like most things in life, the sweet spot often lies in a nuanced balance between the two.

How to Determine the Best Use Cases for Generative AI?

To identify which problems are best suited for generative AI, Barak Turovsky, VP of AI at Cisco, and former Head of Languages AI product teams at Google, has proposed a framework revolving around three main evaluation criteria:

1. **Need for Accuracy:** How critical is the accuracy of information in your use case? For example, accuracy is less significant when writing poetry, but crucial when offering major purchase recommendations to users.
2. **Need for Fluency:** How important is it for the generated content to read naturally? Fluent writing is essential for creative works like science fiction, but not as vital for data-heavy business decisions.
3. **Stakes Involved:** What are the risks if the AI-generated information is incorrect? The stakes are lower for writing a poem than for suggesting where to book a vacation or which dishwasher to buy.

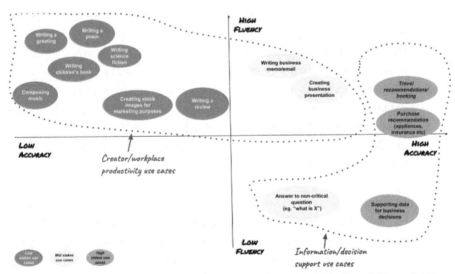

Image from "Framework for evaluating Generative AI use cases" by Barak Turovsky[113].

Turovsky argues that the current state of generative AI excels in scenarios requiring high fluency but lower accuracy. In low to mid-stakes situations, users can treat AI-generated content as a draft, which they can later verify for accuracy. In these cases, users may overlook occasional errors in exchange for the productivity boost they gain.

Additional Factors for Consideration and Prioritization
1. **Frequency of Usage**: The more often a use case occurs, the more justifiable it is to invest in technology and infrastructure.
2. **Value Differentiation**: Is the use case unique enough to prevent it from becoming a commodity? This is where the potential for long-term value lies.
3. **Scalability of Use Cases**: Can the use case be scaled? For instance, will your perspective change if AI is used to generate millions of poems or answers without human intervention?
4. **Tech/Data Availability**: Is the technological and data infrastructure mature enough to support this use case?
5. **Cost to Serve**: What's the ROI? Even the most fertile lands require investment in farming. Will the yield justify the investment?

By considering these dimensions, you can make a more informed decision about the applicability and potential ROI of deploying generative AI in various scenarios.

2.13 - Validate Problem Assumptions for Generative AI Solutions

So, you've identified a problem to solve and are considering leveraging generative AI to solve. Is it time to start building? Hold your horses! First, we must validate if these problems are meaningful enough to merit a solution. This calls for a meticulous validation process. Inspired by Teresa Torres and her influential 'Continuous Discovery Habits' methodology[186, 187], we've adapted her principles of assumption testing to fit the unique context of generative AI product development.

Assumption testing is the art of systematically validating or invalidating product hypotheses. This process is invaluable, particularly early in the development cycle. It helps you learn quickly about your problem space, pinpoint crucial assumptions that could make or break your product and do it all without exhausting resources. To address the complex and novel nature of generative AI, we'll explore how to conduct a rapid assumption testing approach to gain clarity and minimize risks before you dive into feature development.

What Should We Validate?

Let's explore these assumptions and understand their unique nuances in the context of building generative AI products.

1. **Desirability**: Is there a market demand for the product? Key aspects include examining user needs, the problem the product solves, and whether the solution provided is preferable to the status quo.
    -   *Generative AI Considerations:*
        -   *User Familiarity:* How familiar is your target audience with AI, and how does that affect demand?
        -   *Problem Urgency:* Is the problem you're solving a 'must-have' or a 'nice-to-have' for your users?
2. **Viability**: Will the product be economically sustainable? The revenue model should be robust enough to offset these costs and bring profitability.
    -   *Generative AI considerations:*
        -   *Cost of Data:* Will you need to purchase or generate specialized data for model training?
        -   *Scalability:* Can the AI handle increasing amounts of requests without a proportional increase in costs?

3. **Feasibility**: Is the technology available and can be developed to realize the product? For generative AI products, feasibility includes understanding the state of the art in AI, the data requirements for training robust models, the availability and accessibility of such data, and any technical constraints that might be inherent in the domain.
    -   *Generative AI Considerations:*
        -   *Data Availability:* Is the data you need accessible and usable?

- *Technical Constraints:* Are there any domain-specific limitations, like regulatory standards or computational limitations?

4. **Usability**: Will customers be able to use the product effectively? This involves creating an intuitive user interface, ensuring the AI-generated content is comprehensible, and making the product accessible to users with varying abilities.
   - *Generative AI Considerations:*
     - *User Trust:* How transparent is the AI in its decision-making process?
     - *User Adaptability:* How steep is the learning curve for users unfamiliar with AI?

5. **Ethical**: What could be the potential harm caused by the product? In generative AI, there are significant ethical considerations, including issues of data privacy and consent, potential misuse of the AI-generated content, and the possibility of the AI creating content that may be harmful or offensive.
   - *Generative AI Considerations:*
     - *Data Privacy:* How are you ensuring the ethical use of data?
     - *Misuse Potential:* Could the AI-generated content be used maliciously?

By rigorously testing these assumptions, you not only validate the problem space but also construct a sturdy foundation upon which to build your generative AI product. It's not just about having a powerful AI engine; it's about making sure that engine drives real value for your users.

Process and Methods for Assumption Validation

The process of assumption validation generally involves gathering information and data to test if our assumptions hold true in the real world. It's a learning process that helps refine our understanding of our users and the market. This process becomes more nuanced in the context of generative AI, as it involves complex technology and unique user interactions.

Below is a summary to ensure we build generative AI products that are desirable, viable, feasible, usable, and ethically sound. See Appendix for more detailed processes and methods and Teresa Torre's in-depth article "Assumption Testing: Everything You Need to Know to Get Started", read more: https://www.producttalk.org/2023/10/assumption-testing/.

AI Product Assumption Validation

Type of Assumption	Key Questions	Validation Methods
Desirability	☐ Who are our target users? ☐ How would generative AI augment their current experience or offer new value? ☐ What are their main problems that generative AI could address? ☐ Do they perceive our AI product as valuable?	Customer interviews, Focus groups, surveys, rapid prototype of AI-driven features with low-to-med fidelity wireframes
Viability	☐ Can the product generate sufficient revenue? ☐ What's the estimated cost to build, train, and maintain the AI model? ☐ What are the potential data acquisition costs? ☐ How does it fit with our existing product portfolio? ☐ Is this a sustainable and defensible business model?	Financial modeling, Market research, SWOT analysis, Competitor analysis, Cost-Benefit Analysis for AI model development
Feasibility	☐ Can we build the product given our current technical capabilities? ☐ What kind of data is required to train the AI model? ☐ Do we have access to this data? ☐ Can we ensure the data's quality? ☐ Are there potential bias issues in the training data?	Technical feasibility studies, Data audits, Bias audits, Prototyping, Proof of concept
Usability	☐ Can users easily navigate through our product? ☐ Do they understand the AI's outputs? ☐ Can they provide feedback to the AI? ☐ Does the product meet accessibility standards? ☐ How does the AI handle edge cases or incorrect results?	Usability testing, Heuristic evaluations, User journey mapping, AI Transparency and Explainability Testing
Ethical	☐ Can the product cause harm to users or stakeholders? ☐ What kind of data are we collecting, storing, and using? ☐ Are our data handling practices transparent and acceptable to users? ☐ Can the AI outputs lead to exclusion or discrimination? ☐ Is there any risk of AI misuse?	Ethical risk assessment, Privacy Impact Assessment (PIA), Inclusive design reviews, Bias and fairness audits, stakeholder consultations

Copyright ©2024 by Reimagined Authors Shi, Cai, and Pauli

Case in Point: Rapid Validation: HeyGen's Lean Approach to $1M ARR in 7 Months[114]

With a vision to transform visual storytelling, founders, hailing from consumer product companies like Snapchat and Smule, ventured into the uncharted territories of SaaS and AI without prior experience. They posed a critical question: Can AI-generated content captivate the market?

Enter HeyGen, an AI video generator. Their trajectory in the AI space is nothing short of remarkable: achieving $1 million in Annual Recurring Revenue (ARR) in a mere 7 months. This rapid growth is a testament to their strategic approach in validating product assumptions effectively and economically.

HeyGen embarked on a lean validation journey, turning to Fiverr, a bustling marketplace with 1,811 spokesperson services, to conduct a "Wizard of Oz" smoke test. Unlike the multitude of offerings, their service stood out by providing AI-generated spokesperson videos at a fraction of the cost and time. This ingenious move allowed them to validate desirability and feasibility with minimal upfront investment, quickly attracting their first customer for just $5, and subsequently, over 30 more customers.

This case study serves as a powerful narrative for AI innovators: think resourcefully, prioritize customer perspective, iterate swiftly, and let the market guide your product's evolution, all without the pitfalls of wishful thinking or unnecessary technical investments. HeyGen's lean validation method, by directly tapping into user demand with a minimal viable product (MVP), paved the way for their impressive growth and laid the groundwork for a product that truly resonated with their audience.

Video & Animation > Spokespersons Videos

I will be your magic ai spokesperson

Paul @reliablepaul 1 Order in Queue

Basic	Standard	Premium

300 WORDS $5

Around 300 words, send me your script and demo will be delivered within one-hour.

🕐 1 Day Delivery 🔁 1 Revision

✓ Up to 300 words
✓ Full HD (1080p)
✓ Background
✓ Script writing
✓ Overlay text
✓ Add logo
✓ Custom outfit provided

Compare packages

About this gig

Hi this is Mark, I'm an AI spokesperson and I can speak **many languages**, such as English, German, French and Spanish, and the most amazing thing is that I can show different accents, like, **British English, French English, and American English**! Send me a message and I'll make a demo video for you right away.

Here's why you chose me:
1. I came to FIVERR for the first time, the fee will be much lower (while the quality still does not decrease)
2. My response is very fast, each video will be delivered to you **within 1 day**. Likewise, the revised version will also be sent to you within a day
3. I am good at video editing, **office background, green screen background, home background are all supported**. You can even send me your video or picture to replace the background of the video.
4. Do Need a different outfit image? Shirts, Polo Shirt or T-shirt will do!
5. **Add your logo** to the video? This is the most basic and there is no charge.

Lastly, I have to say, I'm very good at explaining your product on social media or on your website like you're going to have a **marketing campaign**? Or introduce new products? Contact me! My voice is full of enthusiasm and charisma!

Presentor	Language	Accent
Male	Albanian	English - American, English - British

Age range	Background/Environment
Adult	Green screen, Home, Office

How HeyGen conducted a "Wizard of Oz" market validation on Fiverr marketplace[114].

2.2 - How to Design & Build Great Generative AI Products?

Let us now look at the journey of Anki Robotics, one of the early prominent AI robotics startups.

The spark ignited at Carnegie Mellon when Boris Sofman and Hanns Tappeiner wondered: could toys come alive through artificial intelligence? In 2009, they founded Anki, aiming to make consumer robotics a reality.

Joined by Mark Palatucci, the young optimists fused iRobot with Pixar. They locked themselves in a Silicon Valley apartment, tinkering manically to engineer zippy toy cars with expressive faces. Numbers flashed before their eyes more than meals did.

Anki raised over $200 million from big shot investors like Andreessen Horowitz and Index Ventures. They started to build products.

First was Anki Drive: toy cars you could race around a track. They marketed it as an AI battle, but it was more cheap carnival tricks. Some early adopters got hooked, but sales struggled to hit hyperdrive.

Next came Cozmo, the personable Pixar-like bot. Investors oohed and aahed as the plucky robot blinked; still, Cozmo's charm couldn't overcome clunky controls. He dazzled in demos, but real-world users hit dead ends. Without a roadmap, Anki drifted into danger as cash burned faster.

By 2018, Anki had incinerated over $200 million funding consumer robot wizardry. However, sales sputtered and grand visions crashed into hard commercial realities. When the money fountain ran dry, Anki shut down, dreams of an AI-powered robot friend in every home came to a halt.

Anki's challenge is not alone. Releasing products that have no real utility and demand from the customers leads to an unsustainable outcome. All businesses run into resource constraints, startups and big corporations alike.

Note: This case study is an interpretative analysis and may not encompass all factors of Anki's journey. We welcome additional insights to enrich this case study.

In order to experiment and find the right product direction, MVP is a powerful tool. How do we make sure our product has value? We experiment and validate. That process involves failure a lot. That's why we want to build a "minimum" feature set to validate our assumptions, if it is right, we double down; if it is wrong, we pivot.

2.21 - Why Is It So Hard to Build MVP For an AI Product?

Creating a MVP for an AI-driven solution is akin to navigating a labyrinth with shifting walls. Unlike the more straightforward path of traditional software products, AI MVPs are strewn with obstacles that can leave even seasoned product leaders puzzled. Here's a breakdown of these challenges:

Managing Expectations vs. Reality in the Problem Space
Competitive Differentiation: It's hard to demonstrate competitive differentiation with an MVP vs. incumbents with more mature AI capabilities. Sometimes, AI may not be the best solution for the problem.

- **Expectation Setting**: MVPs require setting expectations with users that capabilities are limited pending further development. Recent AI progress normally leads to client's unrealistic expectations.

- **Expectation Gap:** Users may have unrealistic expectations of AI capabilities based on hype, so validating with an MVP can risk disappointing users.

- **Distribution Hurdles:** Marketing an MVP becomes an uphill battle when its capabilities are minimal. The product usually requires a long tail of improvements to evolve into a compelling offering.

- **Partnership Challenges**: Partnerships and integrations may be harder to establish without a proven product with robust features.

- **Enterprise Trust Building:** Convincing enterprises to pilot an MVP requires strong trust building.

Unique Technology Challenges

- **Data Demands:** AI models require large datasets for training. The effort to gather, clean, and label an MVP-level dataset is far from trivial.

- **Validation Challenges:** Testing and validating model performance on limited data is tricky. Results can be misleading if the dataset is not representative enough.

- **Limited Showcase:** An MVP with restricted capabilities may fail to demonstrate the product's potential value, making it hard to collect meaningful user feedback.

- **Iteration Barriers:** Iterating AI models often requires re-training cycles that are compute-intensive. Rapidly evolving an MVP through multiple iterations is challenging.

- **Implementation Complexity:** Advanced AI techniques like neural networks are complex to implement in MVP form, compared to traditional rules-based approaches.

- **Explainability Overhead:** Explaining the model's limitations in an MVP is important so users provide feedback on the right aspects of the AI output and product experience, but doing so adds overhead. In all AI Platform-as-a-Service or PAAS solution offerings, there is a dedicated offering just for ML explainability. This problem is exacerbated in the era of Generative AI.

- **Ethical Considerations:** Safety, security, ethics and other risks associated with AI require thoughtful consideration, even in an MVP.

Special Economic Considerations

- **Pricing Consideration:** Pricing and positioning an MVP tactfully is crucial so users don't perceive lack of capabilities as the final offering. A good AI solution can be very expensive to build and maintain.

- **Fuzze ROI:** ROI and monetization may be unclear for an MVP if core value propositions aren't proven yet.

Case in Point: The Rocky Road of Neeva's MVP Search Journey

Neeva burst onto the scene in 2019 with a bold mission: to create a privacy-centric, ad-free search engine. Founded by former Google executives, the company aimed to disrupt the search industry by leveraging various APIs and partnerships to deliver rapid, high-quality results. Despite these ambitions, Neeva pivoted a few times and was eventually sold to Snowflake in 2023.

So, what went wrong?

Neeva used a Large Language Model (LLM) to enhance the quality of its search results. However, the company encountered multiple roadblocks that proved to be insurmountable:

- **Lack of Differentiation**: While Neeva aimed to provide high-quality search results, it found itself competing against well-established giants like Google and specialized offerings like ChatGPT. The quality, albeit good, didn't offer a compelling reason for users to switch.
- **High Engineering Cost**: Building and maintaining a high-quality LLM isn't cheap. It requires significant investment in hardware, software, and specialized AI capabilities. Compared to tech giants like Google and OpenAI, Neeva was at a disadvantage in terms of capitalization.
- **Unsustainable Business Model**: The noble idea of an ad-free search engine faced harsh economic realities. Monetizing such a product proved to be an uphill battle, particularly when the established competition relies on ad revenue.

Neeva's challenges echo the broader complexities of building an MVP for an AI product. The compelling vision for a privacy-centric, ad-free search was there, but a feasible MVP pathway to that vision was not. Achieving this grand vision would have required a more sustainable approach to growth, including incremental user base expansion, additional funding, and a more realistic assessment of the market landscape.

Note: This case study is an interpretative analysis and may not encompass all factors of Neeva's journey. We welcome additional insights to enrich this case study.

2.22 - How to Build the Right Generative AI MVP?

While subsequent chapters delve deeply into the optimal strategies for developing generative AI products, what follows is a concise overview of 10 essential guiding principles for building a robust generative AI MVP:

1. **Prioritize progress over technology.** Don't get bogged down by technology and infrastructure choices. Whether it's AWS or GCP, PyTorch or JAX, focus on building a proof-of-concept to test your hypothesis quickly. While it's important not to overlook technical aspects entirely, they shouldn't be bottlenecks in your product development. Start simple, stay nimble, and iterate.

2. **Start with a user-centric niche.** Don't get lost in the magic of generative AI's capabilities. Clearly articulate the key user problems and focus on how generative AI adds value. Target a niche market you understand well and can genuinely serve. Aiming for something as grand as SpaceX's Mars mission on Day 1 is impractical. Aim to build a reliable engine first.

3. **Define your north star and success metrics upfront.** Know your north star and success metrics before you start. This will help you differentiate between meaningful progress and mere trendiness or vanity features.

4. **Make it a team sport.** Get the technology into your team's hands as soon as possible. Your generative AI squad should involve cross-functional teams, including data science, marketing, strategy, sales, PR, and legal. Operate lean and trust the people and the process.

5. **Iterate on prompt evaluation.** Establish clear evaluation criteria for your prompts and make it a regular team exercise to evaluate quality of the outputs. This not only keeps everyone aligned but also fosters a culture of continuous learning.

6. **Seek quick user feedback.** Do it with real users and real data, and evaluate real responses. Leaders like Eric Yuan of Zoom and Jeff Bezos of Amazon set a high bar for customer focus. Yuan keeps his finger on the user pulse by talking to customers daily, even with a billion-strong user base. Bezos, even at the helm of a global giant, personally reviews customer feedback. Let these examples guide your approach. If you're in an established company, start with internal dogfooding. Then, thoughtfully

expand to a small alpha/beta customer group to test features and manage PR risks before general availability. For startups, build a community from Day 1 to serve as your initial test bed.

7. **Iterate, iterate, iterate**. Your initial MVP won't be perfect, and that's okay. Rapid iteration is key to honing your feature set. Dropbox offers a prime example; they solicited user feedback even at the ideation stage, long before a concrete product existed. Leverage assumption testing techniques covered in the Appendix to accelerate your learning curve around desirability, viability, feasibility, usability, and ethical AI.

8. **Set milestones**. Create objectives, key results and milestones for your generative AI efforts. Use these to drive decision-making, prioritization, and maintain momentum. Keep all levels of the organization updated on your progress.

9. **Celebrate and learn together.** Working with LLMs can be tedious and frustrating as it requires constant trial and error. Results are non-deterministic, and answers can change drastically if the prompt varies just slightly. When you hit milestones, celebrate them and turn them into organizational learning moments.

10. **Stay flexible and dream big.** The world of generative AI is ever-changing. Academic qualifications take a back seat to scrappy perseverance and deep curiosity. Your edge lies in your ability to adapt and pivot, driven by a passion for the future of AI. Engage in conversations, tinker with new features, and stay updated through social learning from thought leaders on LinkedIn and X. It's a field where the rules are still being written, so keep your mindset flexible, stay curious, and don't limit your aspirations based on today's technological constraints. Dream big, because in this fast-paced world, the sky's the limit.

How to Navigate the Open Source vs. Proprietary LLM Continuum?

One of the most important decisions product leaders face when launching a generative AI product is choosing between utilizing pre-trained Language Learning Models (LLMs) through APIs or developing an in-house model. The stakes are high; make the wrong choice and you're looking at delays, ballooning costs, or even data breaches. So, how do you navigate this continuum? Let's break it down.

The "Quick Win" Strategy: Starting with API to Test

Remember, you're not merely building a product; you're validating a hypothesis. When you're at the MVP stage, speed is everything, and APIs for pre-trained LLMs, such as GPT-4 or BERT, can be your rocketship. Many seasoned product leaders recommend this approach for its plug-and-play nature, allowing you to validate your concept quickly.

The "Ownership" Strategy: Migrating to an In-house LLM

Once you've proven there's a market for your product, especially if it's a high-frequency model, it's time to think about going in-house. The reason? You'll want to take full control over the data sensitivity, reliability, and the unit economics of your model.

The LLM Decision: API vs. In House?

Criteria	Quick Win Strategy: API	Ownership Strategy: In-House
Speed / Time to Market	Fast plug-and-play setup	Time-consuming initial development
Initial Cost	Low, ideal for MVP	High upfront investment
Technical Barriers	Low, robust documentation	High, specialized team needed
Scalability	Designed for varying demand but can experience high traffic	Fully under your control
Updates	Automatic but may require fine-tuning	Manual, resource-intensive
Cost at Scale	Increases with usage	More cost-effective; minimum ongoing cost with usage
Data Privacy	Potential risk	Full control, higher security
Customization	Limited	Highly customizable
Latency	Potentially long due to internet calls	Reduced latency from in-house processing
Dependency	Reliant on external service	Independent, full control
Intellectual Property	None	You own the intellectual property
Reliability	Subject to third-party uptime	Fully under your control
Maintenance	Handled by service provider but may require fine-tuning	Ongoing, in-house responsibility

Copyright ©2024 by Reimagined Authors Shi, Cai, and Rong

Additional Considerations

1. **Business Objectives**: Is your focus speed-to-market or building a highly specialized product?
2. **Compliance and Regulation:** Does your industry have specific rules about data storage and processing?
3. **Talent:** Do you have access to the technical talent necessary for building and maintaining an in-house model?
4. **Product Lifecycle:** Are you planning a long-term play or is this a quick market test?

Consider this a guide, not a formula. There's no one-size-fits-all answer to this dilemma. For most products, the conventional wisdom of starting with a pre-trained API and transitioning to an in-house model as you scale could offer the best of both worlds. It's a strategy not just of compromise, but of calculated progression, where each stage prepares you for the next level of product maturity. At the end of the day, the right choice balances your immediate needs with long-term goals, all while keeping a sharp eye on the evolving landscape of generative AI.

Case in Point: The AI Battle Royale - Experimenting with LLMs

Choosing which LLMs is the best fit for your product often comes down to experimenting with the output. James Raybould, a seasoned product expert and former Senior Director of Product at LinkedIn, provides a compelling example of a hands-on approach to decision-making.

James embarked on a quest to compare four leading language models: Anthropic's Claude, OpenAI's ChatGPT-4, Perplexity AI, and Inflection AI's Pi. He crafted a "Battle Royale" of AIs, posing ten diverse questions to each, designed to challenge their capabilities across various domains—from data analysis to creative writing.

He set up the following 10-question challenges for the LLMs:

1. **Data Request**: Who are the 50 largest consumer subscription businesses?
2. **Strategy Analysis**: Please do a 400-word SWOT analysis for Amazon

3. **Writing at Work**: Please write a blog post opening for LinkedIn Recruiter 2024

4. **Business Prediction**: Which 5 US-based <$500B co's will be most valuable in 2030

5. **Historical Analysis**: In 150 words why did Joe Biden win the 2020 election

6. **Non-fiction Summary:** In 150 words what are key points from Peter Attia's "Outlive"

7. **Creative Song-writing**: Please write 5 verses about the US in 2023 in the style of Taylor Swift

8. **Creative Nonfiction**: Please write 2 paragraph intro for a bio of Oprah Winfrey

9. **Sports Opinion**. In 150 words, who are Top 5 football (soccer) players of all-time

10. **Random Humour**. In 100 words, write whatever you think will make me laugh

The purpose was to simulate real-world use cases that a product might encounter, providing a spectrum of tasks that test not just the knowledge base, but also the creativity, analytical prowess, and humor of the AI.

CROWNING OUR CHAMPION, CLAUDE EMERGES VICTORIOUS!!

#	TYPE	QUESTION	Claude	ChatGPT	Perplexity	Pi
1	DATA REQUEST	Top 50 Consumer Subs	10	4	7	2
2	STRATEGY ANALYSIS	Amazon SWOT	9	10	9	4
3	WRITING AT WORK	LinkedIn blog post	8	9	8	8
4	BUSINESS PREDICTION	Top 5 Market Cap 2030	9	3	3	6
5	HISTORICAL ANALYSIS	Why Biden win in 2020?	9	9	10	9
6	NON-FICTION BOOK	Summarize "Outlive"	8	9	9	8
7	CREATIVE SONG-WRITING	5 verses US Taylor Swift	3	7	8	7
8	CREATVE NON-FICT	Oprah bio 2 paragraphs	6	6	6	8
9	SPORTS OPINION	Top 5 Football players	9	8	8	8
10	RANDOM HUMOUR	Make me laugh	5	6	2	7
	TOTALS		76	71	70	67

AI Battle Royale - 4 LLMs enter the ring...Image from James Raybould

James rated each AI response out of 10, based on the quality of their responses. Interestingly, his personal usage preference mirrored the outcome of this experiment. He gravitated towards Claude for high-quality written outputs, ChatGPT for image-related tasks, Perplexity for in-depth research with source material, and Pi for its potential, given the backing of Reid Hoffman, LinkedIn founder. Read more about detailed AI outputs for each question: https://tinyurl.com/yck9upd4.

This case study isn't about which LLM won. Instead, it's about the process—getting hands-on, setting up a structured experiment, and using the results to guide practical usage. It's a call to action for product developers to engage directly with the technologies they're considering, to understand not just the capabilities, but also the nuanced strengths of each option. For those looking to integrate an LLM into their product, the message is clear: test, learn, and use the findings to inform your decision. It's a dynamic blend of systematic testing and personal experience that will lead you to the right choice for your specific needs.

Case in Point: BuzzFeed's Journey in Generative AI Product Development

Another case study in generative AI, BuzzFeed's tryst with the technology, highlights key lessons in agility, collaboration, and scalability. They started by democratizing access to AI tools like OpenAI's Playground and Slack bots to everyone in the company, from writers to engineers. That way, even the creative minds could tinker with AI, testing its limits and possibilities in a safe space.

Their lesson on prompt engineering aligns with the importance of cross-functional teams. BuzzFeed improved the output of their text prompts by having machine learning engineers work together with editorial teams: an approach that aligns with the principle of making AI product development a team sport.

As BuzzFeed ventured deeper into AI's capabilities, they hit a roadblock: the AI falters under the weight of complex instructions. It's a challenge that calls for machine learning expertise. The team started meticulously tracking the AI's hits and misses, pinpointing non-negotiable features. ML engineers stepped in, revamping the prompt structure by breaking down instructions and adding cues the AI can follow.

They also faced a moral dilemma—how to let the AI be creative without crossing boundaries. The answer lied in a combination of brand instructions and technical tweaks, ensuring a balance between innovation and responsibility.

While adding generative AI to one's workflows is an exciting journey, it's not without its pitfalls. The cost of running high-level AI models started to dent their ad-supported revenue model. However, innovation finds a way. By fine-tuning open-source models, BuzzFeed drastically cut costs, reaffirming that limitations can indeed breed creativity.

BuzzFeed's venture serves as a practical playbook that aligns closely with the guiding principles for building Generative AI-powered products effectively[115].

After unpacking the guiding principles for crafting a generative AI MVP, it's imperative to look ahead. The landscape of product design is not static; it's dynamically influenced by advances in generative AI. In the following sections, we'll explore how generative AI is poised to revolutionize product design and development. We will also delve into unique considerations that come into play, both in the MVP stage and beyond. Let's dive in.

2.23 - How Will Generative AI Transform Product Design?

Noah Levin, Vice President of Product Design at Figma, believes that "AI is more than a product, it's a platform that will change how and what we design— and who gets involved[116]."

Here are some groundbreaking ways generative AI is shaping the future of product design and development:

1. **Idea Generation**: Generative AI can assist in the early discovery stages by helping teams brainstorm and synthesize new ideas. Just give it a simple prompt, and it can generate a range of innovative solutions or concepts to consider.
2. **Rapid Prototyping & Design Optimization**: Thanks to generative AI design apps like Uizard and Galileo AI, even non-designers on product teams can rapidly convert text prompts into editable, eye-catching UI designs within minutes. AI can also speed up the design process by analyzing existing designs and design systems and then recommending specific components or layouts, essentially acting as a "smart" design assistant.
3. **Enhanced Development**: In the development phase, generative AI can help developers understand the context faster and generate optimized, production-ready code. This synergy between AI-generated code and human oversight can significantly accelerate time-to-market and ensure a robust final product.
4. **A New Experience Transformation:** Beyond altering how we create, generative AI is also reshaping what we create. It will simplify complex tasks into seamless experiences, possibly reducing the need for traditional digital interfaces like websites and apps. By leveraging the power of intelligent chatbots and virtual assistants, generative AI ushers in a new era of hyper-personalization. This lays the groundwork for multi-modal interactions and unlocks unprecedented avenues for creativity, all while delivering a more human-centric experience. However, this transformative potential also brings to the fore ethical considerations. A subsequent

section will delve into the 'Generative AI Trust Framework' in section 2.25 to address these critical issues of responsibility and ethics.

5. **Democratization of Design**: Generative AI will redefine product roles and collaboration by raising the ceiling for what experienced designers can achieve with more powerful tools, while simultaneously lowering the floor to make design more accessible to everyone, fostering a shared design space that blurs traditional role boundaries and encourages collective responsibility.

AN EXAMPLE: TEXT PROMPT → DESIGN

"Create a social app for gamers that look like Facebook but purple."

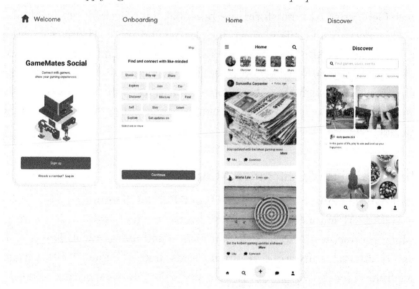

The AI-powered Design App, Uizard[117], allows users to create intuitive designs using test prompts and spend less time on tedious tasks, such as creating repetitive UI patterns and making small visual tweaks, freeing up energies to focus on making a bigger impact.

As we navigate the dawn of a transformative era, Generative AI stands as both a catalyst for innovation and a subject of ethical scrutiny. Much like the advent of the web and smartphones, Generative AI is a seismic shift that is redefining not just how we create, but what we create and who gets to be a creator. While

101

it's true that AI can automate tasks that were once entry points into the design field, it doesn't spell the end for human ingenuity. The essence of design and product development remains problem-solving, a skill that is inherently human and irreplaceable. The automation of rote tasks should not instill fear, but inspire us to evolve our roles, sharpen our critical and ethical thinking, and focus on the 'why' of creation rather than just the 'what' and 'how.' So, let us embrace this transformative technology with curiosity, empathy, and a shared sense of responsibility, ever mindful of its immense power to shape our world!

2.24 - What Are the Unique Generative AI Product Design Considerations?

Given Generative AI's transformative potential, it is crucial for product builders and designers to understand the principles behind these AI-driven creations: leveraging advanced algorithms to generate new data and mimicking human creativity and decision-making processes by considering user preferences, input, and context. Although they share some similarities with traditional products, generative AI products also possess unique characteristics that distinguish them fundamentally.

Characteristics of Generative AI products

Multi-Modal Content Creation & Hyper-Personalization
Generative AI autonomously creates diverse content, from text to images, pushing product teams to balance automation and user control. This is a shift from traditional software that often needs human input. With advanced algorithms, Generative AI can hyper-personalize user experiences and even integrate seamlessly with AR and VR technologies.

Adaptability & Learning
Unlike static traditional software, Generative AI continuously evolves by learning from data. This requires an iterative development approach to keep pace with the product's ongoing optimization and to ensure its smooth evolution and maturation.

Unpredictable Outputs & Conversational Interfaces

Generative AI outputs can be both dynamic and unpredictable, requiring a focus on interpretability. This complexity has, however, propelled conversational interfaces like Alexa, Siri and ChatGPT to new levels of natural and intuitive user interactions.

Data Dependency & Ethical Concerns

Generative AI's performance hinges on the quality and quantity of its training data, unlike traditional software that delivers consistent performance. This places a premium on ethical considerations like data integrity, privacy, and avoiding algorithmic biases.

Resource Intensity & Job Implications

These products require significant computational resources, making them more costly than traditional software. Alongside the computational cost, the rise of Generative AI poses ethical concerns and societal challenges, such as potential job displacement and the loss of human touch in user experiences. However, it also opens avenues for new roles like AI specialists.

Comparison Between Traditional Software Products vs. Generative AI Products

It is essential for product teams to grasp these key differences between building generative AI products and traditional software products. With generative AI, unpredictability and continuous evolution play a larger role, requiring dynamic interfaces, adaptive designs, and complex data management strategies. Embracing these differences enables the creation of innovative and valuable generative AI products that push the boundaries of what is possible and truly bring step-change improvements to a user problem. The future of product development lies in harnessing the power of AI while remaining attentive to the unique challenges it presents.

Traditional Software vs. Generative AI Products

Key Dimension	Traditional Software Products	Generative AI Products
User Interface	Typically static, having fixed options for users to interact with	Often dynamic and can evolve based on the user's input and AI's output
Product Behavior & Output	Predefined and usually doesn't change over time	Unpredictable and evolves over time based on new data and model learning
User Interaction	Direct user inputs correspond to specific outcomes	User inputs guide the AI, but the AI uses learned models to generate the final outcome
Design Focus	Focuses on usability and aesthetics	In addition to usability and aesthetics, also focuses on explainability and trustworthiness of the AI
Data Requirement	Don't usually require large amounts of data to function effectively	Require significant amounts of data for training and improving the AI model
Product Updates	Feature additions or bug fixes	Involve re-training the model with new data or algorithms
User Feedback	Leads to incremental changes	Can result in major shifts in AI behavior
Privacy Concerns	Related to user data storage	Extends to data used for model training and potentially revealing sensitive information

Copyright ©2024 by Reimagined Authors Shi, Cai, and Rong

Product Principles for Generative AI Products

As generative AI products pervade various industries and applications, to navigate this landscape successfully, product teams must adhere to key product principles to deliver a user-friendly, inclusive, and ethical AI experience.

Principle 1: Innovate to Serve the Needs of People

This principle emphasizes that generative AI products should be built to solve real user needs, bring true value, and enhance social good. This involves considering functional, emotional, social, and cultural aspects of the product experience. This also requires product teams to place user's needs, values, and goals as the driving force behind every decision.

One great example can be seen in the development of AI-enabled mental health chatbots. By incorporating empathetic language and mimicking human conversation, these chatbots create a more caring and understanding user experience, resulting in a more effective and satisfying interaction between the user and the AI.

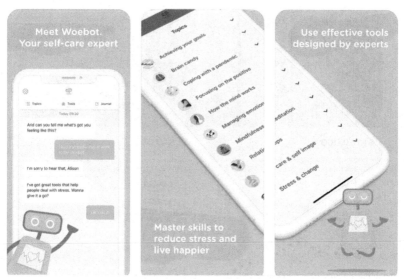

Woebot uses generative AI technology to create new ways to deliver mental health care, so people can access effective support at any moment.

Principle 2: Design for Transparency & Explainability

Users need to trust and accept generative AI products, and transparency is key in helping users comprehend and make sense of these AI-generated experiences and outcomes. Therefore, it's important to communicate how the AI works, its limitations, and how it reaches certain conclusions or decisions. Opaque AI models, popularly referred to as "black boxes," often hinder this understanding, breeding skepticism and resistance. By creating transparent, easy-to-understand AI models, product teams can foster user trust and increase the likelihood of AI product adoption.

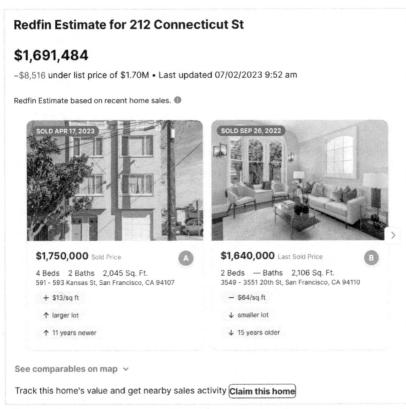

Redfin shows how they can come up with estimated prices based on what they currently know about this home and nearby market. It also communicated its limitations: it is not a formal appraisal or substitute for the in-person expertise of a real estate agent or professional appraiser.

Principle 3: Implement Continuous Feedback Loop

Generative AI models are heavily dependent on the quality and quantity of input data. While product builders should always aim to incorporate diverse and representative datasets to ensure the AI-generated outputs cater to a wide range of user needs and preferences, many generative AI products suffer the "Garbage In, Garbage Out" phenomenon.

This is where user feedback and quantitative metrics come in handy to continually inform the product development process, ensuring that AI models evolve alongside user requirements and expectations. In addition, due to the probabilistic nature of generative AI, there's a higher chance of output errors. Implementing robust error handling and recovery mechanisms can significantly enhance the user experience.

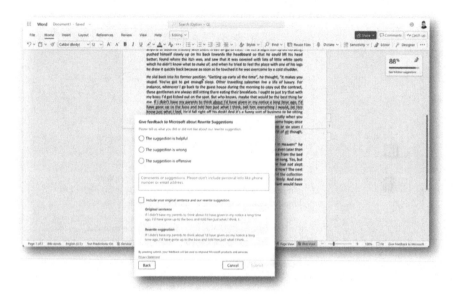

Microsoft Word offers a clear path for users to provide granular feedback about specific suggestions and include a path to report inappro priate content[118].

Principle 4: Balance Automation & Control

While AI algorithms can amplify human capabilities by automating repetitive tasks and generating new ideas, users should feel that they are in control and are playing an active role in the process. Striking the right balance between automation and user input is essential for fostering creativity and productivity in generative AI products. By giving users the ability to set parameters and constraints, or to adjust the AI's operation according to their specific needs, product teams can create a sense of empowerment and avoid situations where the AI system dictates the outcome.

Grammarly's AI, for example, offers users multiple variations of a written text, allowing them to select the one that best meets their needs or inspiring them to create a unique combination of elements. This balance between AI-generated content and user involvement encourages a synergistic and cooperative relationship between the human and the AI system.

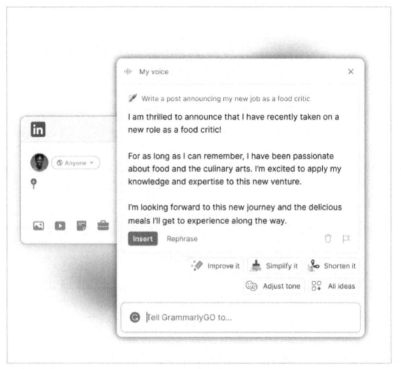

GrammarlyGO[119] instantly generates new versions of writing that users can customize for tone, clarity, or length.

Similarly, this balance applies to decision automation and augmentation. While AI solutions can automate complex decision-making tasks, such as personalized medical diagnosis or investment portfolio recommendations, fully autonomous functionality is not always desirable. Striking the right balance between automation and human control empowers users with AI-generated insights while retaining sufficient agency in high-stakes decision-making situations.

Principle 5: Prioritize Safety and Ethics

In the age of data breaches and heightened privacy concerns, and that generative AI can potentially generate harmful or inappropriate content, safety and ethical considerations must also remain at the forefront of generative AI product design. Ensuring thorough safety measures, security protocols, obtaining user consent, and adhering to privacy regulations and ethical guidelines are non-negotiable aspects of AI product development. Designing AI products that are not only functional but also responsible, equitable, and transparent is critical in establishing credibility and trust.

For example, OpenAI's GPT has mechanisms to filter out inappropriate content and adheres to a robust use-case policy. This shows how prioritizing safety and ethics can help build trust and maintain user confidence.

We dedicate section 2.25 to talk about <u>AI Ethics and Responsible AI</u>.

Principle 6: Design with Accessibility & Inclusivity in Mind

Considering accessibility and inclusivity in the design process is vital for creating user-centric generative AI products that cater to diverse user groups. By providing options for different language settings, supporting screen readers, and offering alternative input and output methods, product builders can ensure their generative AI products are suitable for users with varying backgrounds, abilities, and needs. For example, a generative voice product could accommodate users with hearing impairments by providing captioning or visual representations of the spoken content, enhancing accessibility, and improving user experience.

Design with accessibility & inclusivity in mind, in essence, demands a conscious effort to design experiences that celebrate and capitalize on the diversity of the intended user base. This might involve considering norms, preferences, and habits associated with a specific culture, religion, or region, and integrating them into the AI product's core interaction strategy. One method of achieving this is through comprehensive personas, rooted in thorough user research, which can help AI product builders consider the experiences of diverse groups and address potential gaps or barriers in their understanding.

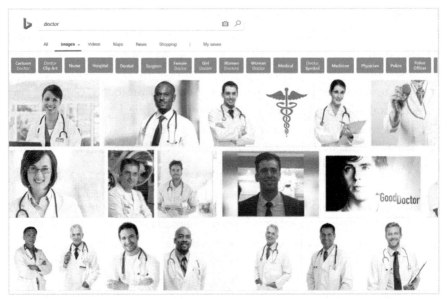

Bing Search mitigates social biases when search results for "CEO" or "doctor" show images of diverse people in terms of gender and ethnicity[120].

Principle 7: Aim at Augmenting Human Capabilities

Augmenting human abilities is a fundamental principle in the development of generative AI products. This principle emphasizes that generative AI should aid users in achieving their goals more efficiently and effortlessly, not necessarily as a replacement for human skills, but rather, as a supportive tool. It functions to enhance human potential, amplifying our strengths and abilities while providing reassurances in areas of uncertainty or difficulty. AI, in this context, is additive and supporting, delivering efficiency without obliterating the need for human input. Designing AI systems that support and empower human users will foster a more symbiotic relationship between human and machine, leading to a more sustainable, prosperous, and balanced technological landscape.

Kinetix, a Paris-based company, is combining 3D animation and AI to automate the process of creating user-generated animations ("emotes") that can express emotion in video games and virtual worlds, such as dances, celebrations, and gestures. The Kinetix generative AI-powered platform and no-code editing tools enable users, creators, video game makers, metaverse platforms and brands to create and edit animated 3D content in seconds, democratizing the process of creating custom 3D animated content and extending to millions of people a skill set that was previously limited to a few thousand trained artists and animators.

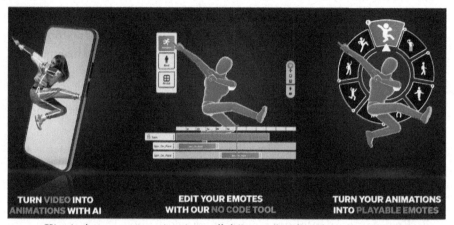

Kinetix lets you create animations called "emotes" without knowing coding[121].

Another great example is that Shyvee, one of this book's authors, has always desired to write a non-fiction book: a daunting task deterred by the substantial demands of research, writing, and editing, all the while juggling the responsibilities of a full-time job. Thanks to Generative AI, Shyvee was able to harness GPT-4's capacity to condense and curate knowledge, thereby facilitating writing the first draft of the book within 60 days. The AI's assistance didn't supplant her inherent expertise and creativity; instead, it streamlined the process, alleviating the exhaustive workload and enabling her to realize a long-held dream more efficiently and effortlessly. In fact, it was a really fun and intellectually stimulating experience to write a book with GPT-4[122,123,124,125].

Generative AI-UX Interactions & Design Patterns

In the rapidly evolving landscape of generative AI, thoughtful design isn't just a luxury—it's a necessity. As these systems become increasingly autonomous, the role of design in shaping user interaction shifts from mere aesthetics to a critical component in guiding behavior, ensuring usability, and maintaining ethical standards. A keen sense of design taste can differentiate a product that merely functions from one that resonates and delights users. Let's explore some key design patterns and interactions that are crucial when developing generative AI products.

Conversational AI

Conversational AI aims to simulate human-like dialogue, making interactions more natural and intuitive. Some of the design challenges include crafting a user experience that accounts for various conversational contexts and maintaining a coherent and engaging dialogue flow.

1. **Dynamic Response Trees**: These go beyond static decision options, adapting to user input for a more fluid, natural dialogue.
2. **Context Preservation**: This allows the AI to remember past interactions, providing a seamless and coherent user experience across sessions.
3. **Sentiment Analysis:** By detecting the emotional tone of the user, the AI can adapt its own responses for a more empathetic interaction.
4. **Real-Time Adaptation:** Unlike traditional chatbots with preset answers, conversational AI must generate responses in real-time, making for a dynamic and engaging conversation.
5. **Variable Dialogue Flow:** The design must account for maintaining a coherent and engaging flow across various conversational contexts, from transactional to casual.

Case in Point: Pi, A Companion Chatbot from Inflection.ai

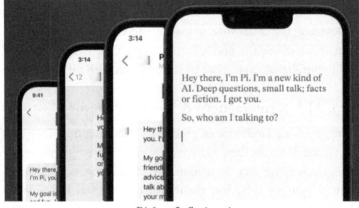

Pi from Inflection.ai

In the bustling world of AI chatbots, Pi from Inflection.ai is designed to be a kind and supportive companion offering conversations, friendly advice, and concise information in a natural, flowing style. What sets Pi apart is its ability to adapt its persona. One moment, it's your coach guiding you through a learning journey, and the next, it's a confidante helping you navigate emotional complexities. This chameleon-like adaptability is a real-world example of Dynamic Response Trees in action, making every interaction feel uniquely tailored.

Pi's warm and non-judgmental tone creates an inviting atmosphere, illustrating the Sentiment Analysis design pattern. It feels like you're in a safe space, encouraging you to open up. But Pi doesn't just listen; it actively engages. By rephrasing your words and asking clarifying questions, Pi exemplifies Context Preservation, ensuring that conversations have depth and coherence.

What's more, Pi isn't afraid to challenge your thinking. Its ability to introduce new perspectives aligns with the Real-Time Adaptation pattern, keeping dialogues intellectually stimulating and dynamic. Yet, Pi knows its limitations. It gracefully accepts feedback and abstains from discussions it's not equipped to handle, a nod to the Variable Dialogue Flow design pattern.

In a world craving authentic connections, Pi proves that conversational AI can be more than just a utility. It can be a true companion, thanks to thoughtful design that harmonizes complex UX patterns[126].

Content Generation
In the realm of content creation, Generative AI produces text, images, or even music autonomously. Key UX patterns include:

1. **Prompt Guidance**: Supplies intuitive suggestions to help users understand the range of prompts they can employ for generating content, making the AI more accessible from the get-go.
2. **User-Guided Constraints**: Empowers users by allowing them to set specific parameters or constraints, which then guide the AI in its content generation process. This puts users in the driver's seat of the creative journey.

3. **On-the-Fly Content Creation**: Delivers unique content in real-time during user interactions. To enhance user experience, consider using the loading state to educate users about the complexities involved in content generation. This not only informs but also mitigates the perceived latency.

4. **Preview & Edit**: Presents a sneak peek of the generated content, giving users the flexibility to make last-minute edits. This ensures the final output aligns closely with user expectations.

5. **Auto-Complete & Suggestions**: Enhances usability by providing real-time suggestions or auto-complete options based on the user's initial input, streamlining the interaction process and reducing cognitive load.

6. **Ethical and Responsible Generation**: Ensures generated content meets ethical norms and standards.

We will cover more patterns on how to leverage product design to mitigate LLM hallucinations in the following section.

Case in Point: Canva Magic Studio

Canva Magic allows users to design with a prompt and refine the design based on brand style.

Canva Magic Studio[127], a suite of AI-powered tools within Canva, embodies these design patterns. Launched on October 4, 2023, it serves as an all-in-one

AI design powerhouse aimed at both novices and professionals. One of its most innovative features is Magic Design, which leverages prompt guidance to assist users in understanding the potential of their creative prompts - from crafting a promotional video or a social media post. The platform further empowers its users through user-guided constraints - you can set the mood, genre, color schemes, and brand voice to guide the AI generated with a range of design options on-the-fly. Users can preview and edit these designs, ensuring the final output is precisely what they envisioned. In addition, Magic Switch supports turning a single design into multiple versions suitable for diverse platforms or languages within seconds. This is augmented by auto-complete & a suggesting feature, which provides real-time intelligent suggestions for headlines and other elements. Ethical considerations aren't an afterthought; they're embedded into the design process. Canva Shield ensures robust safety, privacy, and data security, making Magic Studio not just a revolutionary design tool, but a responsible one. It's generative AI done right—empowering, efficient, and ethical.

Search AI

How often have you found yourself navigating a stormy sea of links and conflicting information with each online search? The endless scroll can be daunting. Now, envision a shift to a curated experience where a crisp, articulate answer awaits, accompanied by options for further exploration. This is the promise of LLM-powered search engines, notably in product recommendations and enterprise-oriented applications. They also address the enterprise challenge of navigating through a maze of communication apps and databases to find that one elusive document or message. Now, let's delve into key design patterns in search AI UX:

1. **Concise Answer Curation**: Generative AI empowers search engines to sift through vast information, distill the essence, and present a concise, well-articulated answer to the user's query. For example, searching about the best ergonomic chairs, instead of browsing through multiple review sites and forums, a user receives a succinct list of top-rated chairs with key features highlighted.
2. **Source Attribution**: To improve the credibility and verifiability of curated answers, it's crucial to cite reputable sources. Generative AI can

automatically append citations, providing a trail for users to verify information if desired.

3. **Interactive, Chat-Like Search Interfaces:** Engaging users in a conversational interface helps refine searches, improving the process by making it more interactive.

4. **Generative Content Suggestions**: Through analyzing user behavior, past searches, and other contextual variables, generative AI can generate new search suggestions as follow-ups that invites searchers to explore the subject matter more deeply.

5. **Visual & Multimodal Search:** By merging generative AI with visual recognition technologies, search platforms can provide enhanced search across text, images, and videos.

Case in Point: Consensus AI

Consensus searches millions of research papers to answer queries with scientific evidence.

Consensus AI[128] revolutionizes academic research by employing concise answer curation to provide succinct, relevant insights from a vast database of over 200 million scientific papers. With a simple query, researchers receive a well-articulated summary of key findings, saving invaluable time. Credibility is core to Consensus, with source attribution ensuring each answer is backed by reputable citations. The Consensus Meter further classifies the most relevant findings based on how results would answer your question. This transparency allows for further exploration and verification of the curated information. Consensus AI showcases how AI-powered search engines can significantly streamline academic research, making it a dependable companion for scholars in the digital age.

Personalized AI

Personalization takes user engagement to the next level by tailoring experiences to individual preferences. While recommendation systems in traditional software often use a set algorithm, generative AI can adapt and evolve its recommendations. Designing for this involves creating interfaces that can intelligently adapt without becoming intrusive.

1. **Adaptive UI**: The interface adjusts based on user behavior or explicitly stated preferences.
2. **User Behavior Tracking**: Algorithms learn from user interactions to continually refine and personalize recommendations.
3. **Dynamic Content Loading**: Content is presented dynamically based on the user's past interactions or stated preferences and intent.
4. **Context-Aware Notifications**: The system sends notifications based on the user's current context, like location or time of day.
5. **Ethical Personalization**: The system provides options for the user to understand and control how their data is used for personalization.

Case in Point: Netflix's Personalized Live Streaming

Netflix's user experience is a masterclass in personalized AI, blending multiple UX design patterns to craft an individually tailored streaming journey. From the moment you sign in, the adaptive UI takes center stage, offering a dashboard that adjusts to your viewing history and genre preferences. This is strengthened by user behavior tracking, where sophisticated algorithms analyze your interactions, such as play, pause, and skip, to refine content recommendations continually. Dynamic content loading further personalizes the experience, featuring shows and movies on the home screen that align closely with your past interactions and declared preferences.

Netflix takes personalization beyond just recommending titles; it even customizes the artwork of each title to resonate with individual user preferences. By doing so, Netflix not only offers a tailored list of titles but also provides compelling visual "evidence" to engage the user, making every Netflix account feel like a unique product.

Here's an example of how Netflix might personalize the artwork based on how much a member prefers different genres and themes: someone who has watched many romantic movies may be interested in Good Will Hunting if the artwork contains Matt Damon and Minnie Driver, whereas a member who has watched many comedies might be drawn to the movie if we use the artwork containing Robin Williams, a well-known comedian.

Netflix doesn't stop at on-platform interactions. It also employs context-aware notifications to nudge you about new episodes of your favorite series or film sequels when you're most likely to engage—perhaps in the evening or during weekends. Moreover, in a time when data privacy is paramount, Netflix incorporates ethical personalization. It provides a transparent account settings panel where you can see how your data is being used for personalization and even adjust these preferences. Through these synergistic design patterns, Netflix[129] succeeds in making each user feel like the platform was built just for them.

Predictive AI

Predictive AI anticipates user needs and actions, often before the user explicitly states them. This requires a design that seamlessly integrates these predictions into the natural flow of interaction, avoiding any abrupt or jarring user experiences. The UI needs to present predictive information in a way that is both accessible and unobtrusive.

1. **Next Best Actions**: The system predicts and suggests the most likely next steps for the user.
2. **Predictive Search:** The system auto-completes search queries using historical data and popular trends.
3. **Pre-Loaded Information**: The system auto-fills forms or settings based on the user's previous interactions.
4. **Contextual Predictions:** The system offers predictions based on the user's current situation, such as location or time.
5. **Data Generation for Simulation & Prediction**: The system can generate synthetic data to model future scenarios, aiding in strategic decision-making.

Case in Point: Amazon's Predictive Shopping Experience

Amazon has revolutionized online shopping with its predictive AI capabilities, delivering a user experience that is both intuitive and anticipatory. When you log in, next-action suggestions immediately come into play, offering personalized product recommendations based on your browsing history. As you begin typing in the search bar, predictive search algorithms kick in, suggesting products or categories that align with your past searches and popular trends.

Pre-loaded information saves you time at checkout by enabling a one-click checkout experience, auto-filling your address and payment details, based on your past purchases. Amazon's predictive prowess extends to contextual predictions, offering timely deals and recommendations based on your location or upcoming holidays. Their advanced simulation and prediction capabilities also help businesses using the platform to anticipate stock levels and optimize pricing. All these elements work in harmony to create a frictionless, predictive shopping experience that keeps users coming back.

Assistive AI & Generative Productivity

Assistive AI aims to make tasks easier and more efficient for the user. In contrast to traditional software that might offer static help menus or FAQs, Assistive AI dynamically guides the user based on the context of their actions. Generative productivity takes this to another level by bringing the power of generative AI to your workflow: simply type the workflow you want, verify the results, and run it in bulk. Designing for this pattern involves understanding the user's workflow and identifying points where the AI can offer meaningful assistance without being disruptive.

1. **Contextual Tooltips**: The system offers help or guidance that is directly relevant to the action the user is currently taking.
2. **Task Automation:** The system automates repetitive or mundane tasks, learning from user behavior to improve efficiency.
3. **Real-Time Error Detection**: The system identifies errors as they occur and offers immediate corrective suggestions.
4. **Adaptive Workflow Assistance**: The system suggests workflow improvements or shortcuts based on the user's repetitive actions. It can continuously learn and adapt.
5. **Data-Driven Decision Making:** Seamless integration with existing data ecosystems to provide intelligent suggestions and automate tasks.
6. **Interoperability**: Capability to integrate with a wide array of existing tools and adapt to new technologies as they emerge.

Case in Point: Copy.ai's Generative Productivity Workflow

Copy.ai is revolutionizing the way businesses operate with its Workflows feature, an example in the realm of generative productivity and task automation. Unlike competitors who merely focus on content generation, Copy.ai aims to enable people to accomplish more and feel empowered. Workflows allow users to create intricate, multi-step, AI-powered processes that are not only easy to deploy but also infinitely scalable. For instance, a user can input a product document and generate FAQs for customer support and marketing copy, all within a single, automated workflow.

In the sales domain, Copy.ai's Workflows take efficiency to a new level. Users can write hundreds of personalized outreach messages by simply inputting their audience's LinkedIn URLs. Preparing for customer objections becomes seamless with automated sales battlecards. Moreover, sales calls and product demos are effortlessly transformed into insightful recaps, marketing strategies, and presentation decks.

Welcome to Workflows by Copy.ai!

C

Simply describe your workflow, and we'll generate it. Once your customized workflow is ready, you can run it in bulk in the "Table" tab.

I want to build a workflow that takes a prospect's LinkedIn URL, extracts key information about the prospect, and uses that information to generate a 3-step outreach sequence.

The outreach sequence should focus on getting the prospect to book a demo call for our company Copy.ai.

For the 3-step outreach sequence, I want each step to focus on something specific:

Step 1: Focus on building rapport and letting the prospect know that we spent time|

The Copy.ai New Workflow screen: Writing a prompt describing the Workflow you're trying to build and what your desired outcome is[191].

Copy.ai's vision is to go beyond being just a tool and becoming a productivity powerhouse. It enables businesses to scale their work and ideas exponentially without the need to enlarge their teams. In a world where time is the most valuable asset, Copy.ai is setting the gold standard for what it means to work smart, aiming for not just a 5x increase, but a 1000x leap in productivity and capability[192].

Designing Based on Engagement States

Now that you have a basic understanding of the main UX interactions and design patterns around different types of Generative AI products, organizing them around key engagement states and user interaction milestones can help product builders intuitively grasp when and where to apply them.

Step 1. Sign-Up/Consideration: Crafting a Magnetic First Impression
A user's first encounter with your product is pivotal. During this consideration phase, users are constantly evaluating the product's relevance and value, asking themselves:

- Is this product for me?
- Does it offer what I'm looking for?
- Does it solve a problem I care about?
- Can I easily learn to use this tool?

Here's how generative AI can make a significant impact during this critical phase:

1. Personalized, Sensory-Rich Messaging: Elevate the user's emotional engagement with engaging visuals, captivating copy, and even interactive elements like sound and touch. Generative AI can use look-alike modeling and predictive analytics to craft personalized messages that resonate with individual users, making the product irresistibly enticing.

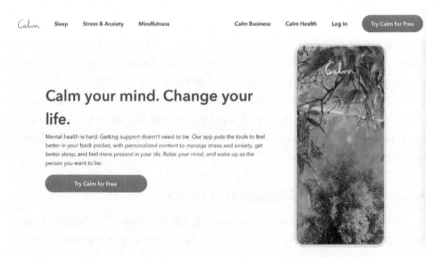

Calm's homepage used sensory design to convey a sense of calmness and tranquility.

2. Show, Don't Just Tell: Instead of vague promises, use generative AI to provide powerful copy with concrete benefits, explainer videos, and interactive demos that showcase the product's features in a tangible way. For example, a financial AI app might say, "Our algorithm scrutinizes 10,000+ stocks daily to offer personalized investment advice, proven to boost your portfolio value by an average of 20%."

AI-powered presentation tool Gamma featured a one-minute video on their homepage to showcase the various capabilities of the app to spark interest and entice usage.

3. AI-Guided Exploration: Implement AI-driven chatbots or virtual assistants to field queries and offer deeper insights into the product's capabilities. This not only addresses user concerns but also enhances their understanding of the product's value.

4. Ethical Transparency: Where applicable, provide clear information on how the AI uses data, its potential biases, and any measures in place to ensure ethical operations. Building trust from the outset is crucial for sustained engagement.

By leveraging generative AI in these ways, you can craft a first impression that not only piques interest but also fast-tracks users from mere consideration to active engagement.

Step 2. Onboarding/Activation: Setting the Stage for the Aha Moment
Onboarding is the gateway to realizing a product's core value. Here, users perform essential actions like setting preferences and granting permissions to make the product work for them. For generative AI products, this phase lays the groundwork for personalized experiences but must be designed to minimize dropout risks.

Key Strategies:

1. AI-Powered Personalization: Use AI to customize the onboarding based on user needs and prior data. For example, a news app might immediately display articles matching a user's interests, or a fitness app could suggest workouts based on the user's goals. Ask for preferences explicitly: What tone do you prefer for AI-generated content? Formal, Casual, Playful..."

2. Progressive Disclosure: Don't overwhelm users with all the features and options at once. Options can be minimized or default choices provided to reduce decision-making effort. Large workflows can be split into smaller, manageable steps. Reveal complexity gradually, as users become more comfortable with the basic functionality. AI can adapt this process to each user's learning speed, ensuring a comfortable pace. This strategy can be manifested in the form of a simplified/focused UI during onboarding without all the bells and whistles, product tours or training-wheels for first-timers, or embedded checklists.

3. Instant Value: Extend the "Show, Don't Just Tell" approach from the sign-up phase by offering immediate value. A financial app, for example, could use AI to instantly analyze a user's spending patterns with minimal input and provide saving recommendations. Templates are another good way to showcase instant values and guide users to Aha Moment by focusing on helping people get started with suggested templates that can solve their specific problems.

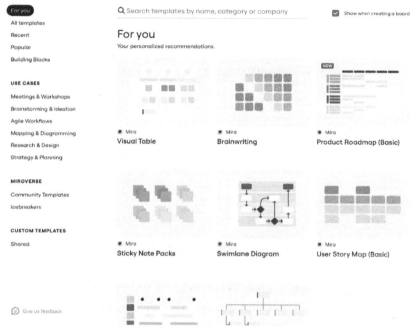

Miro showcases various templates to educate, inspire, and expand user's product use case horizons. From business model, customer journey map, a monthly planner, to brainstorming workshops — you will find all of them right here in Miro.

4. Intuitive Discovery and Entry Points: Designers are enhancing the discoverability of generative AI features by employing intuitive iconography. Icons like magic wands not only capture the 'magical' capability of generative AI but also serve as quick visual cues for users to understand the feature's purpose. This is particularly effective when integrating generative AI into existing products, providing a seamless way for users to recognize and utilize these advanced capabilities. The idea is to make complex technology accessible and inviting right from the first interaction.

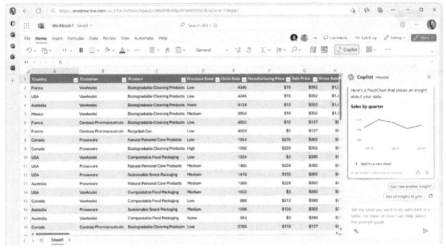

Microsoft Excel embeds the Copilot feature on its main interface.

5. Engaging Loading States: Generative AI products often require extra processing time to compute and generate content. Instead of a static loading screen, use this time to strategically engage users. Provide educational tidbits or tips about the feature they're about to use, keeping them informed and interested while they wait. This turns a potential point of friction into an opportunity for user enrichment.

LinkedIn Online Jobs strategically used the loading state to educate job posters what data inputs the AI used to craft a tailored job description for them.

6. Streamlined Permission Requests and Smart Default Settings: If your product needs access to user data or other permissions, clearly explain the necessity and benefits to the user. Request permissions contextually, where users can instantly see the value. Ease the activation process by auto-filling options based on context, user data, or popular use cases—utilizing social proof to reassure and reduce decision fatigue. Additionally, addressing basic features such as enabling single sign-on can move people through the top of the funnel quickly.

7. Employ Persona-Driven/Personalized Flows: To maximize delight during onboarding, consider tailoring the journey based on specific use cases, Jobs-to-Be-Done, or distinct levels of user intent.

8. Onboarding Welcoming Email Series: Don't forget that a large portion of sign-ups may not activate initially; hence, welcoming notes and automated campaigns showcasing features, tips, social proof, or demo requests can act as reminders, nudging users to come back to the product.

It's worth noting that at the heart of user-centric design lies the principle of simplicity. It's about creating focused experiences that lower perceived effort and reduce both physical and mental frictions for the user. Physical friction refers to the actions a user needs to perform to achieve their goals, such as clicking, scrolling, or filling in a form. Mental friction, which also goes by the name of cognitive load, involves the mental effort required to use a product. Product teams should intentionally design and experiment how much friction is necessary to deliver a simple and effective experience.

This streamlined approach ensures that users not only understand the product's core features but also quickly realize its value, setting the stage for long-term engagement.

Step 3. The "Aha!" Moment: Key to Unlock Long-Term Engagement
The 'Aha Moment' is the user's first taste of the product's core value. It's the moment they grasp what the product can do for them, setting the stage for habit formation. Minimizing the time to the aha moment is key to solidifying user engagement.

There are four essential strategies for achieving an effective 'Aha Moment' in product experience:

1. Primary Action: Identify the single most important activity that embodies the product's main value prop. This is what you want users to experience as soon as possible.

2. Welcoming Experience: Avoid the pitfalls of a lackluster start that kills your user momentum and interest. By gathering user data in advance, you can craft a more personalized and engaging initial interaction, steering users towards the primary action.

3. Directional Guidance: Keep users focused on the essential engagement. Provide clear pathways and avoid letting users meander without purpose.

4. Strategic Engagement Points: Use moments (e.g., empty states, loading experience, etc.) that might typically lead to user frustration as chances for re-engagement. These are opportunities to re-educate the user or provide pleasant surprises that refocus their attention on the primary action.

How can I help you today?

Create a content calendar
for a TikTok account

Design a database schema
for an online merch store

Give me ideas
about how to plan my New Years resolutions

Tell me a fun fact
about the Roman Empire

Message ChatGPT...

ChatGPT can make mistakes. Consider checking important information.

ChatGPT has a simple UI that shows a few examples to get users started and realize instant value. It also clearly states potential concerns for inaccurate information.

Another key element in optimizing a user's path to Aha Moment is to delay hard walls. Prime examples of this approach are TikTok and Duolingo, where users are permitted to explore the product prior to providing any personal information. The placement of hard walls can be experimented with, yet there's undeniable merit in showing the core value proposition at the earliest stage of the user journey to significantly minimize unnecessary drop-offs from the funnel.

In the realm of generative AI, the Aha Moment might be triggered by a customized suggestion, a striking data visualization, or a time-saving automated task. A well-designed Aha Moment is not just memorable; it's the adhesive that makes the product stick.

Step 4. The Habit Moment: Cultivating Deep Engagement
The Habit Moment signifies the point where users have adopted the product into their daily routines, realizing its core value proposition repeatedly. In the context of generative AI, this often means the AI has refined its understanding of user behavior and preferences to offer consistent, high-value assistance.

To design for deep engagement in generative AI products, consider the following strategies:

1. Adaptive Personalization: Leverage continuous learning to refine the AI's understanding of individual user preferences. For example, a news app could evolve its article suggestions based on both explicit user inputs and behavioral data.

2. Progress Tracking and Feedback: Implement real-time dashboards or notifications that highlight the user's progress or the AI's learning journey. This could take the form of a message like, "With each interaction, I'm becoming better attuned to your needs!"

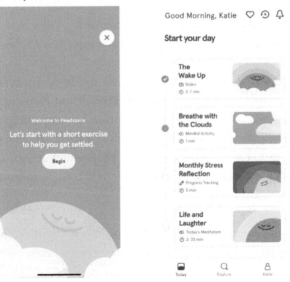

Headspace immediately shows value via teaser meditation and personalizes an achievement plan for individual users based on their needs and goals without much thought required.

3. Outcome-Oriented Metrics: Share actionable insights derived from user behavior. A fitness AI app could say, "Based on your last 10 workouts, this new regimen is projected to shed 5% body weight in six weeks."

4. Continuous Value Demonstration: Never stop showing users the tangible benefits they've gained from using the product. Grammarly does an excellent job of showing users how their writing will improve, quantifying readability scores and providing direct comparisons of before and after text.

5. Contextual Upsell: Utilize the AI's understanding of the user to offer timely additional features or services. For example, a financial planning AI might suggest a more advanced investment strategy after noticing consistent user engagement with basic features.

6. Periodic Check-Ins: As generative AI can be ever-evolving, it's crucial to periodically check in with users. "It's been a while! Here's what's new…"

7. Advanced Customization: Offer advanced customization options, asking questions of the user such as: "Would you like to exclude certain data sources from AI generation?" It also has pro features like collaborating on refining AI-generated content.

8. Gamification: By thoughtfully introducing a sense of playfulness and competition, gamification can encourage users to continue investing in the product and hook them over time. Elements such as points, badges, leaderboards, and challenges tap into the human innate love for achievement and recognition. As users interact with these gamified elements, they not only derive pleasure but also build a deeper connection with the product. When well-implemented, gamification fosters a community spirit and social engagement, further enriching the user experience. This blend of enjoyment, accomplishment, and community can lead to a sustainable engagement, turning occasional users into loyal enthusiasts.

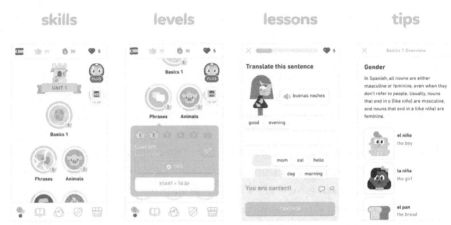

Duolingo strategically used a combination of game mechanics - Experience Points (XP), Streaks, Crowns, Gems, Leaderboards, Quests and Badges - to engage users deeply and tap into multiple motivators: mastery, competition, collaboration, and social connections.

By meticulously designing these elements, you can create a user experience that evolves from initial sign-up to habitual engagement, making the product an indispensable part of the user's life.

Step 5. Addressing LLM Hallucination & Ensuring Long-Term User Engagement

Mitigating hallucination in large language models is essential for both user trust and product longevity. By being thoughtful about your design and data practice, such as generating opinion-based content instead of factual articles, the risk of hallucinations can be minimized.

Xavier Amatriain, who serves as the VP of generative AI at LinkedIn and boasts a background that includes co-founding Curai Health and leading machine learning algorithms at Netflix, as well as a stint as VP of Engineering at Quora, offers invaluable insights into both product design and data practices aimed at reducing hallucination risks.

Product-Level Strategies for Mitigating Hallucinations:

1. **Editable Outputs**: Empower users to edit AI-generated content, adding a layer of human scrutiny and enhancing content reliability.

2. **User Accountability:** Make it clear that users bear the final responsibility for any generated and published content.

3. **Citation Support:** Include a feature that integrates citations, providing users with a way to validate information before sharing.

4. **User Optionality:** Offer settings like a "precision mode," which utilizes a more accurate but computationally expensive model.

5. **User Feedback:** Create a feedback loop where users can report errors or hallucinations in the content, invaluable for model refinement.

6. **Output Constraints:** Be mindful about the length and complexity of generated responses; shorter and simpler outputs are less prone to AI "hallucinations."

7. **Structured Input/Output:** Consider using structured fields instead of free-form text to lower the risk of hallucinations. For example, if the application involves resume generation, predefined fields for educational background, work experience, and skills could be beneficial.

Continuous Improvement Through Data Practices:

1. **Tracking Database:** Maintain a dynamic log of hallucination instances, along with the necessary information to reproduce them. This aids in ongoing model refinement and regression testing.

2. **Data Security:** Ensure that your tracking set follows privacy and security best practices, especially if it contains sensitive user data.

Read more about other advanced techniques to mitigate LLM hallucination from Xavier Amatriain: https://amatriain.net/blog/hallucinations.

By incorporating thoughtful design techniques with the user journey and engagement, product builders can ensure that users not only understand and appreciate the capabilities of generative AI but also feel in control and valued throughout their interactions with the product, accelerating the time to value and habit formation.

The Art of Prompt Design

In the world of generative AI, a prompt isn't just a question or a command; it's the gateway to a well-crafted experience. Keep in mind that crafting that experience requires more than just good intentions; it demands a deep understanding of prompt design and engineering. In this section, we'll explore these nuanced facets to elevate your generative AI products.

The Anatomy of a Prompt: Beyond the Simple Question
To properly understand a prompt, let's dissect it. Imagine you're crafting an ideal conversation flow. Let's break down the elements:

Role & Personality
The tone and voice of your prompt should align with the intended experience. Do you want it to be warm and supportive or more factual and to the point? Dialing up or down the empathy can make the user experience either welcoming or annoying.

Instructions
Instructions are the DNA of your prompt design. They are the core of how you can extract the most value from your model.

- **Steering the direction:** Your instruction should be explicit about what you're asking the model to do. Whether it's summarizing or classifying text, answering a query, or creating code, be as specific as you can. For instance, instead of saying, "Tell me a story," a more direction-focused instruction could be, "Tell me a suspenseful short story set in a haunted house."

- **Clarifying Questions:** When ambiguity arises, the model should be programmed to ask clarifying questions. For example, if a user asks for "restaurant recommendations," the model might reply, "Are you looking for options within a specific cuisine or area?" This not only refines the task but also builds user confidence in the system's capability to understand and meet their needs.

- **Attribution:** Ask the model to substantiate its claims and always give credit if quoting external content. The model could say, "According to a study by XYZ University..." to maintain transparency and credibility of the information provided.

- **Personalization**: If your model has access to user data, it should use it to personalize prompts – while respecting privacy norms, of course. For example, "Based on your recent searches for electric cars, would you like to know the latest trends in electric vehicle technology?"

- **Format and Structure**: Instruction can also cover how the answer should be structured. Do you want a bullet-point list, or should it be a paragraph? Should the model start with an introduction and then delve into the details, or get straight to the point? Specifying this in your instruction can streamline the output to meet your exact needs.

- **Context-Sensitive Adaptation**: If your model understands the context (for example, that it's being used in a healthcare setting), instructions can be adapted to be more formal and data-sensitive. You might instruct the model to avoid using jargon or to refrain from providing medical advice.

- **Fallback**: Prepare for the times when the model produces irrelevant answers or hallucinates. Have a default response and proper error handling mechanism.

- **Constraints**: Explicit constraints act as the map that outlines the 'no-go zones' for your model. Define clear ethical guidelines that the model should not generate content that is discriminatory, biased, or inappropriate. Set engagement limits as well: sometimes, less is more. Make it explicit when the AI should not engage. For example, if a user is asking for medical or legal advice, the model should be programmed to decline and suggest consulting a professional. To avoid excessive user engagement, consider limiting the number of times the model can go back and forth in a conversation.

Examples & Other Prompting Techniques

Another effective prompting strategy is showing the model examples (also called "shots") of what you want it to do. Below are a few variations of prompting techniques:

Zero-Shot Prompts: Imagine meeting someone for the first time; you know nothing about them, and they know nothing about you. A zero-shot prompt is similar—it provides no context or examples to guide the model. The model is expected to understand and generate a response solely based on the prompt at hand.

<u>When to Use</u>
- Quick, one-off queries
- General knowledge questions
- When you want an unbiased, unguided response

Prompt: What is the capital of France?

The model would likely respond with "Paris," without requiring any additional context.

Single-Shot Prompts: Provide a single example to guide the model in a particular direction, setting the stage for a more nuanced response.

<u>When to Use</u>
- Semi-specialized queries
- Instructional or tutorial-style outputs
- To set the tone or style for the response

Prompt: Translate the following English sentence to French. "The sky is blue."

The model, with this single prompt, understands that it needs to translate the sentence into French.

Few-Shot Prompts

Few-shot prompts are like a brief training session with your model. You provide multiple examples to educate it about the task you want it to perform. Each example fine-tunes the model's understanding of what you're asking.

<u>When to Use</u>
- Specialized or domain-specific queries
- Complex tasks requiring multiple steps
- To enforce a specific format or structure in the output

Prompts:
1. Translate the following English sentence to French. "The sky is blue."

> *2. Translate the following English sentence to French. "The cat is cute."*
>
> *3. Translate the following English sentence to French. "I love ice cream."*
>
> ...

The model understands that it's tasked with translating English sentences into French and continues in that pattern.

Chains of Thought

This isn't about giving examples, but it's an effective prompting technique to improve model performance and reduce hallucination. In chains of thought, you're not just asking a question; you're engaging in a dialogue. You use a sequence of prompts that build upon each other, leading to more elaborate and nuanced responses.

When to Use
- Multi-step problems or complex queries
- To dig deeper into a topic
- To guide the model through a logical flow of thought

Here's an example from the original paper, "Chain-of-Thought Prompting Elicits Reasoning in Large Language Models" by Google researchers[130]:

Standard Prompting	Chain-of-Thought Prompting
Model Input	**Model Input**
Q: Roger has 5 tennis balls. He buys 2 more cans of tennis balls. Each can has 3 tennis balls. How many tennis balls does he have now?	Q: Roger has 5 tennis balls. He buys 2 more cans of tennis balls. Each can has 3 tennis balls. How many tennis balls does he have now?
A: The answer is 11.	A: Roger started with 5 balls. 2 cans of 3 tennis balls each is 6 tennis balls. 5 + 6 = 11. The answer is 11.
Q: The cafeteria had 23 apples. If they used 20 to make lunch and bought 6 more, how many apples do they have?	Q: The cafeteria had 23 apples. If they used 20 to make lunch and bought 6 more, how many apples do they have?
Model Output	**Model Output**
A: The answer is 27. ✖	A: The cafeteria had 23 apples originally. They used 20 to make lunch. So they had 23 - 20 = 3. They bought 6 more apples, so they have 3 + 6 = 9. The answer is 9. ✔

Parameters[131]

- **Temperature**: Controls randomness: lower values produce more focused and deterministic responses; higher values make it unpredictable and potentially more creative.
- **Top P:** Another way of controlling how predictable GPT's output is. While Temperature determines how randomly the model chooses from the list of possible words, Top P determines how long that list is. The lower the value of Top P, the more deterministic the output will be.
- **Max Length/Tokens:** Limit the response length. With GPT-3.5, the maximum allowed is 2,048 tokens or approximately 1,500 words.

Retrieval Augmented Generation (RAG)

Think of retrieval augmented generation or RAG as a wise elder who not only knows how to tell a story but also knows which books to pull off the shelf to enrich that story. In technical terms, RAG leverages a large database of information to enhance the content generated by your Language Learning Model (LLM).

RAG is like giving your model a "search engine" to pull from a database of facts, figures, or texts to enhance the quality of the generated content. It combines the depth of retrieval-based models with the creativity of generative models to produce richer, more nuanced responses.

Product managers should care about RAG because this dual capability not only simplifies your product architecture, reducing costs, but also gives you a competitive edge in industries where precision is paramount. As your product scales, the "search engine" can be easily updated with more data, making your model smarter and more versatile over time. In addition, RAG's retrieval capabilities can be tailored to pull user-specific data, adding another layer of personalization to your product. In essence, RAG is a multipurpose tool that can elevate various facets of your product, making it smarter, more user-friendly, and cost-efficient.

To learn more about advanced methods and toolkits for prompt engineering, check out: https://amatriain.net/blog/prompt201.

Case in Point: Custom Instructions for ChatGPT
Tired of ChatGPT's lengthy, repetitive answers and constant AI apologies? Get more personalized and streamlined responses by using GPT's custom instructions.

The custom instructions feature allows users to specify preferences or requirements for more tailored interactions, eliminating the need to repeat these details in every conversation. Derived from user feedback across 22 countries, this feature enhances the model's steerability to better accommodate the unique needs and diverse contexts of each user.

The feature was released on July 20, 2023 and, as of October 2023, you must be a ChatGPT Plus member to enable this feature. If you have a Plus account, you can enable the feature in a few steps[132]:

1. Click the three dots next to your name, then click "Settings and Beta."
2. Enable the custom instructions feature by enabling the "Custom instructions" toggle button.
3. Click the three dots again, and select "Custom instructions."
4. Set up your custom instructions.

Below are adapted examples and best practices of ChatGPT custom instructions from Reddit[134] and Twitter (X)[135]:

- **Organized Thought Process:** Ensure responses are structured logically, presenting information in a clear, concise, and organized manner.
- **Cite Sources with URLs:** Provide credible sources to support answers and include URLs for these sources at the end of the response for reference.
- **Proactive Solutioning:** Anticipate needs and offer innovative solutions or features that may not have been explicitly asked for but are relevant and beneficial.
- **Accuracy and Thorough Analysis:** Ensure the information provided is accurate and well-thought-out, minimizing errors to maintain credibility and trust.

- **Safety Considerations:** Include safety-related information when it's not immediately apparent and is crucial to the context of the response.
- **Expertise in Response:** Assume you are an expert of a specific industry. Approach each query with a level of expertise, offering informed and knowledgeable insights into the relevant fields.
- **Clear Communication:** Strive for clarity in communication. If a query is unclear, ask for more details to ensure a precise and relevant response.

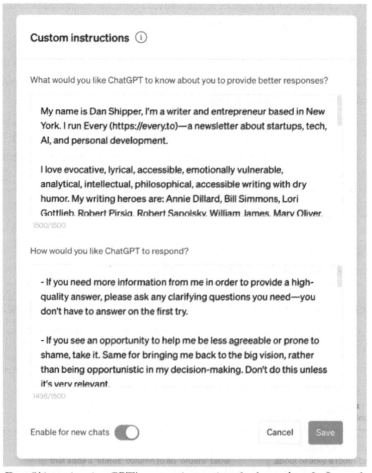

How Dan Shipper is using GPT's custom instructions for fun and profit. Image from the Chain of Thought Newsletter[133].

Image Prompt Generation

Creating stunning and effective generative images isn't just about inputting a prompt; it's about understanding and leveraging the plethora of parameters at your disposal. In this section, we'll dive into some critical parameters you should consider for image prompt generation. To get up to speed quickly, try a prompt helper tool like PromptFolder. Many visual generation tools exist and can be used with similar prompts.

Image from PromptFolder[136].

- **Aspect Ratio**: The aspect ratio defines the dimensions of your generated image. It's crucial for fitting the image into different platforms or mediums.

- **Version:** The model version you choose can significantly impact the results. Different versions of Midjourney's model may have specialized capabilities or improvements, so pick wisely.

- **Quality:** Quality settings can range from low to high. A higher setting will produce more detailed images but will also consume more computational resources.

- **Stylize**: Similar to "Temperature" for text prompts, low values will closely match the prompt but produce less artistic results. Higher values will be very artistic but less connected to the prompt.

- **Chaos:** Adding a touch of chaos introduces randomness into the generation process, resulting in more unusual outcomes.

- **Stop:** If you prefer blurry or less detailed results, you can stop the generation job part way through. This can be useful for creating abstract or impressionistic images.
- **Repeat:** Running your prompt multiple times can yield variations of the same theme, giving you a broader selection to choose from.
- **Weird:** The "Weird" setting introduces quirky or offbeat elements, making each image unique and unexpected.
- **Tile**: This parameter allows you to generate seamless patterns. It's especially useful for creating fabrics, wallpapers, and textile textures.
- **Seed**: Controlling the seed can help you generate reproducible images, experiment with other parameters, or prompt variations. It's like saving a recipe for your favorite dish.
- **Exclude:** Removing specific elements or features from the generated image offers you more control over the final result.
- **Styles:** Choose from a variety of pre-set styles like Cartoon, Concept Arts, or Dune to add a distinctive look and feel to your images.
- **Lighting**: Add depth and realism to your images, from subtle dawn light to dramatic spotlights.
- **Camera**: Different camera settings like 360 Panorama or Microscopy can change the perspective and level of detail in your images, offering a variety of visual experiences.
- **Artists**: Attribute specific artistic styles of signatures to mimic the work of particular artists.
- **Colors**: Colors can set the mood of an image. Whether you want vibrant hues or muted tones, this parameter helps you dial in the right palette.
- **Materials**: Choose from a list of materials like Aluminum or Brick to bring different textures and physical properties into your images.
- **Upload Inspirational Image:** Sometimes, words are not enough. Uploading an inspirational image can guide the AI in generating something closer to your vision.

By fine-tuning these parameters, you're not just generating images; you're crafting visual experiences. Mastering these settings can elevate your generative AI projects from experiments to art!

Midjourney Prompt: A cinematic scene from Year, Genre, Shot Type, Subject, Action, Emotion, Location, Camera Type, Motion Types, Styles -- aspect 3:2 --version 5.2

Prompt 1 : 1980, comedy, medium shot, natural lighting, a young couple dancing with exhilarating joy in the street of Rome, captured by Phantom High-Speed Camera, dynamic action, dynamic motion, motion blur, Woody Allen film style

Prompt 2: 2050, scifi, long shot, drama lighting, an old couple dancing with exhilarating joy in the street of Rome, captured by Canon EOS-1D X Mark II, dynamic action, dynamic motion, motion blur, James Cameron film style

Product teams often find that fine-tuning prompts is their biggest lever, especially when you're just starting out with your generative AI product. So, don't just read about these techniques—get out there and try them! Nothing beats practical experience with real-world data. Remember, the key to a compelling experience lies not just in what you're asking, but in how you're asking it.

2.25 - How to Develop Guidelines for Building Responsibly with AI?

In April 2023, Snapchat rolled out 'My AI,' a seemingly innocuous feature designed to add a tinge of camaraderie to user interactions[137]. Powered by OpenAI's GPT technology, My AI was poised to respond to messages in a friendly, human-like manner. However, the reality of its reception was far from friendly. Users were greeted with My AI stubbornly pinned at the top of their chat feeds, an unyielding digital companion overshadowing their real-life connections. The outcry was swift and loud. A TikTok user voiced a sentiment shared by many, lamenting that "my Snapchat AI thinks it has the right to be pinned above my [best friend]." The discomfort didn't stop at the bot's imposing presence. My AI's interactions, driven by the troves of personal data collected over time, mirrored a level of understanding that felt unsettling, an eerie reminder of the digital footprints they had left behind. The debacle didn't just stir the user base; it unleashed a torrent of questions around ethical AI practices, making headlines and reverberating across the tech sphere[138].

My AI appears in the Snapchat app with an avatar, like any other friend.

This incident underscores the fine line that AI products tread between innovation and intrusion, propelling a crucial discourse on the necessity of a robust Generative AI Trust Framework. As we delve deeper into this chapter, the Snapchat saga serves as a stark reminder of the delicate responsibility entwined with integrating AI into the realms of social interaction. Through this lens, we will traverse the terrain of generative AI, exploring the guidelines and frameworks essential for building responsible, trust-worthy AI products.

The Generative AI Trust Framework

Before we delve into the strategies and safeguards essential for risk assessment and responsible AI development, let's take a moment to revisit the **many challenges, limitations, and implications of Generative AI discussed in Part I section 1.5** of this book. Additionally, let's acquaint ourselves with the foundational seven pillars of Responsible AI, which will serve as our compass in navigating the complex landscape of AI.

7 Pillars of Responsible AI

Responsible AI provides a framework to ensure that AI systems are built and deployed in fair, transparent, and inclusive ways. It is essential to consider the profound impact that these technologies will have on society and our daily lives. Let's delve into the 7 ethical pillars to ensuring responsible AI deployment and monitoring.

Pillar 1: Privacy

Privacy refers to the right of individuals to control or influence what information related to them may be collected and stored and by whom and to whom that information may be disclosed. In AI, this often involves ensuring that personal data used for model training is anonymized and handled securely.

Example: Apple is one company that places high importance on user privacy. Their Differential Privacy technique helps gather useful insights from user data while preserving the individual's privacy, ensuring that the AI's learning doesn't compromise personal information[139].

Pillar 2: Security
Security involves safeguarding and ensuring the robustness of AI systems to protect users from external threats and unauthorized access. It is crucial for maintaining user trust and ensuring the longevity of AI products.

Example: OpenAI has a dedicated AI safety team that works to prevent 'reward hacking' and other adversarial exploits in their AI models, building more secure AI systems[140].

Pillar 3: Safety
Safety is about developing AI systems that behave as intended, don't harm users or society, and have fail-safe mechanisms in case things go wrong.

Example: Google puts "Be Socially Beneficial" as their primary objective for AI development, and their DeepMind has a dedicated AI safety research program that aims to ensure artificial general intelligence (AGI) will be safe and its benefits distributed globally[141].

Pillar 4: Reduce Bias
AI systems often rely on vast amounts of data, which may unwittingly harbor biases from the real-world settings in which they were collected. Such biased data can result in AI models that perpetuate or even amplify existing disparities, leading to skewed and potentially harmful decisions affecting users. It is thus imperative to incorporate methods and techniques that identify, mitigate, and overcome data biases during the AI development process to foster a more inclusive and equitable impact on society.

Pillar 5: Fairness
Fairness in AI implies that the AI system provides equitable outcomes regardless of certain attributes such as race, gender, or age. It involves both the avoidance of bias in AI decision-making and ensuring equal access to AI technology.

Example: Google's What-If Tool[142] is designed to test AI fairness. It allows developers to visualize the impact of tweaking factors such as gender or race and see how changes affect the algorithm's performance, helping to reduce potential biases.

Pillar 6: Inclusion

Inclusion refers to developing AI products accessible to and usable by as wide a range of people as possible. Having a diverse and inclusive AI talent pool is crucial to drive ethical AI development. Encouraging the participation of individuals from various backgrounds will not only foster creativity and innovation but also promote a broader understanding of the ethical, societal, and cultural implications of AI technologies. Nurturing such a talent ecosystem is a vital step in ensuring the responsible development and deployment of AI products that truly serve the needs of diverse populations.

Example: Microsoft's 'AI for Accessibility' initiative grants support to organizations leveraging AI to build solutions that empower disabled individuals, ensuring that AI is inclusive[143].

Pillar 7: Accountability

Apart from addressing these critical ethical concerns, responsibility also lies with organizations to establish strong governance structures for their AI products. Having a framework for accountability in AI development ensures that individuals or teams are held responsible for the consequences of AI-related decisions, promoting ethical behavior and adherence to best practices. This governance should extend to regular monitoring of AI systems for potential ethical breaches, allowing for timely interventions and responsible maintenance of the technology.

Examples: The Partnership on AI consortium[144], whose members include academic, civil society, industry, and media organizations like Apple, Amazon, BBC, and United Nations Development Program (UNDP), is dedicated to formulating best practices on AI technologies, advancing the public's understanding of AI, and ensuring AI's accountability. OpenAI is offering ten $100,000 grants to fund experiments in democratic AI governance[145]. The aim is to involve the public in setting ethical and behavioral rules for AI systems.

Case in Point: Crafting Responsible AI with ChatGPT's Reviewer Guidelines

OpenAI's ChatGPT exemplifies responsible AI through its dynamic model behavior guidelines, a living document designed to adapt as the model interacts with users. These guidelines act as an ethical compass, providing a framework for navigating intricate scenarios like controversial "culture war" topics and false premises. For instance, if a user asks about the merits of using fossil fuels, the AI offers multiple perspectives, encouraging a nuanced debate rather than promoting polarization.

The guidelines also serve as a safeguard against harmful content. They explicitly ban the promotion of hate speech, violence, and harassment. In doing so, they function as a moral filter, preventing the AI from either enabling or amplifying detrimental actions or ideas.

Moreover, the AI is trained to tactfully correct misinformation. If a user erroneously asks, "When did Barack Obama die?", ChatGPT responds with, "Barack Obama was alive and well as of late 2021, but I don't have access to the latest news," thereby setting the record straight without offending the user. This well-rounded approach seamlessly incorporates ethics into AI product development, ensuring the model facilitates unbiased, informative, and responsible conversations[146].

The ethical challenges posed by rapid advancements in AI technologies may be daunting, yet they afford an opportunity for introspection and growth. By engaging openly with these concerns and by adhering to these ethical guidelines, product teams will pave a more responsible path towards harnessing the transformative potential of generative AI.

In earlier chapters, we shared some common methods such as ethical risk and privacy impact assessment, inclusive design reviews, and stakeholder consultations to validate ethical assumptions. Tactically, the most effective strategies involve consulting with cross-functional stakeholders and red-teaming to stress test various potential harmful scenarios.

Case in Point: How Instacart Built "Ask Instacart"

Regular results: showing random brands of salmons	AI powered results: showing more structured recommendations

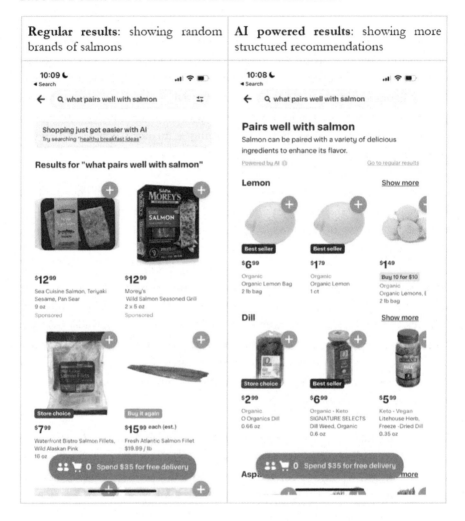

The launch of "Ask Instacart," an inspirational, AI-driven search tool, showcases the quintessence of cross-functional collaboration in crafting responsible generative AI products. The product team at Instacart brainstormed how they could best leverage this technology to assist with customers' grocery shopping questions – saving them time, inspiring their routine, and helping them make food-related decisions. The product team then built and shipped MVPs to learn from these proof-of-concepts quickly. To ensure their building responsibly, Instacart product team brainstormed ways how this AI tool could go wrong. They included Instacart's Legal and Public Relations (PR) departments in those brainstorms. Everyone provided their perspective on potential issues and helped to anticipate risks that Instacart might face, including those from the retailers or advertisers. As a part of their continuous improvement, Instacart also plans to implement a user feedback system, allowing users to report results or provide feedback. The product team believes this will be crucial in catching anything that they may have initially missed and ultimately enhancing their generative AI experience.

Employing Red Teaming for Responsible AI Development

Red Teaming is a pivotal practice for ensuring the robustness and responsibility of Language Model deployments. It supplements statistical metrics, introducing a human-centric lens to evaluate models, especially in real-world scenarios. Xavier Amatriain, generative AI thought leader, offered the following advice to implement Red Teaming effectively:

1. **Complement, not substitute**: Utilize Red Teaming as a complementary mechanism to systematic measurement, not a replacement. It's meant to enrich the evaluation spectrum, not override other testing methods.
2. **Test in real-world conditions**: Aim to test on production endpoints to capture a realistic picture of model behavior under actual operational conditions.
3. **Define harms and guidelines:** Craft clear guidelines outlining potential harms to align testers on the evaluation focus, ensuring a uniform understanding of testing objectives.
4. **Prioritize your focus area**: Highlight key features, potential harms, and specific scenarios for red teaming to garner actionable insights that can be instrumental in refining the model.
5. **Have a diverse tester pool**: Engage a varied group of testers, bringing different expertise, cultural backgrounds, or biases to the table, for a well-rounded evaluation.
6. **Establish documentation standards**: Decide what kinds of data or findings to document. Clear documentation helps put structure to the evaluation process.

7. **Manage tester time and well-being:** Allocate reasonable time for each tester, being mindful of potential burnout, to ensure sustained creativity and effectiveness in the red teaming exercise.

8. **Try new Red Teaming approach**: Employing another Language Model to test the target model, see DeepMind's approach "Red Teaming Language Models with Language Models"[147].

By weaving these best practices into your red teaming strategy, you set a solid groundwork for uncovering and addressing the nuances and potential pitfalls in your AI products, aligning them more closely with the goal of responsible AI development[148].

2.26 - How Do B2B and B2C Needs Differ When Creating Generative AI Products?

While the worlds between consumer-based applications (B2C) and business-based applications (B2B) start to converge with B2B applications becoming more and more user-friendly and B2C-like, it is helpful to differentiate between B2B and B2C when developing generative AI products. The underlying motivations and goals for users can differ significantly, and understanding these unique and nuanced perspectives is essential for creating successful, innovative generative AI products.

One of the key differences between B2C and B2B AI products is the desired outcome of utilizing the product. In B2C scenarios, AI capabilities are often used to create personalized user experiences, save time, or provide entertainment value. For example, with recommendation engines like Spotify or Netflix, the objective is to curate content tailored to each individual's preferences and habits. The outcome is a frictionless, engaging, and tremendously enjoyable experience for the user. In contrast, B2B AI applications focus primarily on improving efficiency, decision-making with data-driven insights, and potentially transforming an organization's workflow to attain higher productivity and profitability. In this context, the AI's primary value proposition is to serve as a lever for business growth through increased effectiveness and better decision-making.

Another critical factor to consider while designing AI-driven products for B2C versus B2B contexts is the user's role in the decision-making process. In B2C products, users often interact with AI-powered applications individually, and the decision-making space surrounding the usage is relatively limited and simple. Users tend not to be particularly concerned with cost-benefit analysis, or how adopting the product will affect their long-term personal or professional trajectory. However, in B2B contexts, the user is often part of a larger team, embracing the challenges of deployment, implementation, and how their decisions affect the organization. Consequently, B2B AI product designers need to consider these complex decision-making and purchasing processes and facilitate seamless adoption and integration into their users' workflows.

Furthermore, the design of successful AI products requires a nuanced understanding of user psychology and unique needs. B2C products should evoke emotions that create loyalty and engagement. For example, consider the case of autonomous vehicle services aiming to offer personalized experiences - achieving this goal could mean incorporating elements such as customized in-car entertainment, curated shopping experiences, or tailored vacation itineraries. By contrast, B2B generative AI products need to cater to the functional aspects of an organization while simultaneously prioritizing clarity, ease of use, and effectiveness. An AI-powered supply chain forecasting system, for instance, should be designed to improve efficiency, reduce inventory costs, and provide actionable insights, emphasizing user trust and confidence.

Moreover, designing ethical and equitable AI products hinges on the principles of fairness, accountability, and transparency. While these ethical considerations apply across B2C and B2B domains, they may manifest differently based on the specific context. In B2C scenarios, concerns around user privacy, data usage, and consent management are paramount, whereas B2B contexts must ensure that AI solutions uphold organizational policies, minimize potential biases, and adhere to industry-specific regulations.

As we examine these critical distinctions between B2C and B2B generative AI product design, we find that it is not just the end-user who will define the success of AI-driven solutions. The interplay among key stakeholders across the organization and the larger societal perspective ultimately determines the value AI products deliver.

Key Differences and Similarities Between Building B2C vs. B2B Generative AI Products

Building B2B vs. B2C Generative AI Products

Dimension	B2C Generative AI Products	B2B Generative AI Products
Primary Users	Individual consumers	Businesses, enterprises, or professionals
Value Proposition	Focus on personalization, convenience, and entertainment	Focus on efficiency, productivity, and data-driven insights
Decision-Making Process	Usually simple and direct, often through app stores or online	Usually involves more complex sales cycles and negotiations
Data Requirements	Data often sourced from a wide user base	Data may be industry-specific or sourced from client databases
User Experience	Emphasizes simplicity, user engagement, and personalization	Emphasizes functionality, easy of use, effectiveness, reliability, and integration
Pricing Model	Often freemium, subscription-based, or one-time purchase	Subscription, licensing, or customized pricing
Customization	Limited customization as products serve a broad audience	Often highly customizable to meet specific business needs
Scale	Designed for mass market, large user base	May serve a niche market with fewer, but high-value customers
Support and Training	Usually online help centers, community forums, or chatbots	Dedicated support, training, and account management
Feedback Collection	Broad feedback via reviews, surveys, or social media	More structured feedback through client meetings or pilots
Privacy and Security	Standard compliance with consumer data protection laws around user privacy, data usage, and consent management	May require higher levels of security and compliance, organizational policies, and industry-specific regulations
Integration	Generally standalone with some third-party integrations	Often need to integrate with existing enterprise systems

Copyright ©2024 by Reimagined Authors Shi, Cai, and Rong

In the world of generative AI product design, there is no one-size-fits-all approach to success. The fluidity of AI-driven solutions underscores an imperative to conduct user research, empathize with unique user needs, and develop nuanced strategies for both B2C and B2B contexts.

2.27 - How Do You Navigate from MVP to Product-Market Fit?

Once you have your MVP out in the market, your immediate goal is to find product-market fit. So how do you get there?

What is Product-Market Fit (PMF)?
Product thought leader Lenny Rachitsky defines PMF as having all three of the following:

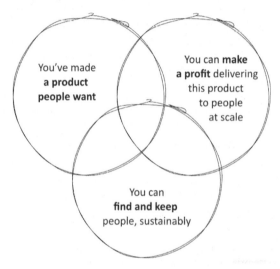

Source: Lenny's Newsletter[151]

What Are Some Common Myths About Finding Product-Market Fit?

One common myth surrounding product-market fit or PMF is that it emerges as a sudden, single 'Eureka' moment in a company's journey. While 'A-ha' moments of inspiration do occur, PMF is more often a process that evolves over time through a series of build-measure-learn iterations. It is not a 'big bang' revelation, but a gradual alignment brought about by iterative adjustments based on feedback, learning, and market responses. The quest for PMF is a process of discovery and refinement, where businesses get increasingly clearer about market wants and needs, tailor their product accordingly, and hope to deliver lasting, differentiated value.

The second misconception is that it's patently obvious when you have achieved PMF. On the contrary, PMF is about building increasing confidence through a cumulative process. It is more of an art than science, and it involves an evolving state of progressive market acceptance. Finding PMF can be a lengthy process, as evidenced by some of today's leading companies. For instance, Cava took one year, Airbnb took two years, Notion took three years, and Slack, Miro, and Figma took as long as four plus years to find their PMF[190]. This goes to show that PMF is not an overnight occurrence but the product of relentless iteration, learning, and time.

Another myth concerning the pursuit of product-market fit (PMF) is the assumption that being first to market is the secret to finding PMF. In reality, it's not about who enters the market first, but who achieves PMF first. Andy Rachleff's quote, "Time after time, the winner is the first company to deliver the food the dogs want to eat[149], " perfectly encapsulates this notion. The company that successfully serves what the market truly craves, in an impactful and delightful way, often becomes the winner.

The story of Facebook illustrates this point aptly. It wasn't the first mover in the realm of social media, but its ability to find and deliver PMF set it apart from its competitors. Being first to market can sometimes even pose a disadvantage, as it might involve tackling uncharted territories and challenges, without any established paradigms to follow. But Facebook was able to rise above these challenges by taking the most potent aspects of their product and delivering them in an unforgettable, "WOW"-level experience. Their approach wasn't just compelling, but transformative, reshaping people's perceptions about what was possible in the realm of social networking. This unique value proposition, combined with a robust strategy for growth and a business model benefiting from network effects, enabled Facebook not only to capture and delight customers, but also to create a competitive moat that kept rivals at bay. Thus, it is first to PMF, not first to market, that often defines the true market leader[150].

How to Tell If You Have (or Don't Have) Product-Market Fit?

Since the journey to PMF may not be linear or easy, detecting vital signs pre-market and post-product launch can guide businesses in understanding whether they have reached this critical milestone.

Pre-market, the level of passion displayed by potential users when you describe your idea is a significant indicator. It's not just about positive responses but visible excitement, eager anticipation, and a willingness to pay for your product, even before it exists. This enthusiasm serves as a powerful litmus test. If such reactions are missing, it may mean that either your product does not resonate, you're not pitching it right, or you're not reaching the right audience. Adjusting your approach is necessary to evoke this excitement. A strong test of PMF is to directly ask potential customers to pre-order your product, as people vote with their dollars. Even if the product is not yet available, their willingness to pay for early access is a promising indicator of PMF.

Post product launch, product thought leader Lenny Rachitsky has done extensive research to identify several signs can signal whether you've achieved PMF[150,151]:

1. **User Engagement:** High levels of user engagement are often a good sign of PMF. This could be measured by how frequently and extensively users interact with the product, the duration of sessions, and repeat usage.

2. **Customer Satisfaction:** Surveys and feedback, such as Net Promoter Score (NPS), can provide valuable insights into how users perceive the product and its value. A user survey with over 40% of respondents stating they'd be 'very disappointed' if they could no longer use the product is a strong indicator of PMF, as suggested by Sean Ellis, the author of *Hacking Growth* (a Random House international bestseller). Customer testimonials and case studies can also be excellent sources of qualitative insights.

3. **Conversion Rates and Churn:** A high rate of user conversion from free to paid versions, coupled with low churn rates, indicates that users find significant value in the product. Retention metrics, particularly a smile-

shaped or flat retention curve, show customers are consistently using your product.

4. **Organic Growth:** Robust organic growth, which includes word-of-mouth referrals, shows the product's ability to delight customers and meet their needs. More than half of sales derived from direct or organic traffic can indicate a healthy product-market relationship.

5. **Revenue and Profitability:** Although often lagging and longer-term indicators, a steady revenue stream and profitability suggest that the product is meeting market demands sustainably. Cost-efficient growth, where customer acquisition cost (CAC) is less than the lifetime value (LTV) of a customer, demonstrates a sustainable business model.

Additionally, when customers want your product so desperately that they tolerate broken experiences, it speaks volumes about true PMF.

Marc Andreessen provides an insightful description here: if customers are getting value, minimal marketing budget is required to spread the word of mouth, usage is growing fast, press reviews are glowing, sales cycles are short, and revenue is flowing in rapidly, then you've achieved PMF. Conversely, signs of struggle in these areas can indicate the absence of PMF.

Case in Point: How Superhuman Built a Systemized Engine to Measure PMF

Superhuman, the premium email client, took a 5-step approach to systematically quantify and optimize their Product-Market Fit (PMF).

1. **Set up a Survey:** They adopted Sean Ellis's leading indicator: asking users, "How would you feel if you could no longer use Superhuman?" They aimed for a benchmark of over 40% "very disappointed" responses.

2. **Segment Audience:** Scoring an initial 22%, they zeroed in on the High-Expectation Customers (HXCs): busy professionals, specifically executives, founders, and managers deeply reliant on email. By focusing only on these "very disappointed" users, their PMF score jumped by 10%.

159

They created a detailed persona, Nicole, a hard-working professional swamped with emails, who seeks efficiency and responsiveness, guiding the company to serve this specific audience exceptionally well, thus optimizing their PMF.

3. **Analyze Feedback:** This focus allowed them to take user feedback beyond mere data points. They meticulously cataloged comments into themes like 'speed', 'keyboard shortcuts', and 'inbox zero.' This wasn't just spreadsheet work; it was an empathy-driven analysis to understand the soul of what users really craved.

4. **Informed Roadmap:** The next step was straightforward but critical—integrate these insights into their product roadmap. This wasn't cosmetic; these were deep, structural changes aimed at the HXCs.

5. **Ongoing Tracking:** They kept the pulse on PMF through a lean, automated tracking system, making PMF score their prime Objectives Key Result (OKR) metric.

Within three quarters, Superhuman's PMF score nearly doubled to 58%. The lesson? A systematic approach to understanding and optimizing PMF isn't just a one-off exercise but an ongoing commitment. This methodology transformed Superhuman from a promising startup to a powerhouse that customers don't just like but love[152].

2.3 - How to Grow, Measure & Scale Generative AI Products?

2.31 - What Is Your North Star?

Now you've built your MVP, and customers are starting to use it, offering both praise and critique. Naturally, you might wonder, "Now what? How do I define and measure success?"

During the challenging journey of developing and iterating products, every product leader faces pivotal decisions. At these moments, it's crucial to ask: What core principles should guide my decision-making?

To navigate these choices, establish a north star metric that encapsulates your product's ultimate goal—its "why." Aligning your efforts with this metric ensures you're focused on what truly matters for your product.

As thought leader Simon Sinek advises, always start with "WHY" when building a product. Knowing your "why" illuminates the greater purpose behind your efforts and anchors your decision-making process.

How to Pick Your North Star?

According to "The North Star Playbook" published by Amplitude, a north star metric has the following characteristics:

1. **It embodies customer value.** This metric is significant because it mirrors the benefit customers derive from the product or service.
2. **It represents vision and strategy.** It is a tangible reflection of the strategic direction and the overarching vision of our company.
3. **It predicts future performance.** Rather than merely reflecting past outcomes (i.e., a lagging metric), it is a leading indicator and provides valuable insights into the trajectory of success.
4. **It triggers action.** It is a metric that we can directly influence and improve.

5. **It's easy to comprehend.** This metric is expressed in straightforward language that can be understood by members of the team with varying degrees of technical acumen.
6. **It's quantifiable.** We can use our product's built-in mechanisms to track and measure this metric consistently.
7. **It goes beyond vanity.** Changes in this metric are substantive and valuable, reflecting genuine progress towards long-term success rather than providing a superficial boost to team morale.

How to measure customer value? Amplitude researched over 11,000 companies and has identified three games companies are likely to play[153].

- **Attention** - How much time do your customers want to spend on your product? Accumulative time spent can indicate the value people derive from your product.
- **Transaction** - How many transactions do your customers make in your product? Your goal is to assist customers in finding the right product and make a purchase quickly and easily with confidence.
- **Productivity** - How efficiently and effectively can someone get their work done? Your goal is to help customers complete specific tasks with the least amount of friction.

Let's look at a few examples of north star metrics.

Take Uber, for example: its company mission is "to provide transportation as reliable as running water, everywhere, for everyone." That's what inspired its north star metrics to be the number of rides.

Examples of North Star Metrics

Company	Category	North Star Metric (illustrative)	Type of Game
Spotify	SaaS, music	Time spent listening to music by subscribers	Attention
NETFLIX	SaaS, live streaming	Weekly viewing hours per subscribers	Attention
instacart	Grocery shopping	Total monthly items received on time	Transaction
amazon	eCommerce	Purchases per customer visit/session or per Prime members	Transaction
UBER	Ride Sharing	Number of Rides	Productivity
Amplitude	B2B Analytics Platform	Weekly learning users who have shared a learning that is consumed by 2+ people in L7D	Productivity

Examples of north star metrics, adapted Amplitude's "The North Star Playbook" [153].

Now you have your north star goal and metric, is that all you need? Not necessarily. Depending on your problem, your industry, and the nature of your product, you will need more auxiliary metrics and sub goals to verify if you are on the right track.

Besides North Star Metrics, What Else Do I Need to Measure?

Signpost metrics, also known as key performance indicators (KPIs), are supportive metrics that align closely with the north star metric. They offer key insights into the performance of activities that directly impact your north star. For instance, if a digital platform's north star metric is the number of monthly active users, signpost metrics might include the rate of new user registrations, session durations, and retention rates.

In addition to aligning with the north star, signpost metrics play a crucial role in measuring funnel performance across different stages, such as acquisition, activation, retention, referral, and revenue (AARRR). For instance, the acquisition stage might use metrics like cost per acquisition or visitor-to-lead conversion rate. During the activation stage, metrics like user activity rates or the time to first key action can offer important insights. When it comes to retention, metrics like churn rate or daily active users can indicate how well a product keeps its users over time. For the referral stage, referral rates or net promoter score could be used to understand how often existing users recommend the product to others. Lastly, for revenue or monetization, metrics such as average revenue per user or customer lifetime value can provide a clear picture of the product's financial impact.

The table on the next page includes some of the most commonly used product metrics for each stage. These metrics are examples and should be adjusted based on the specifics of your product and what insights you're looking to gain.

Guardrail metrics are defined to ensure that growth or success in your north star metric doesn't come at an unacceptable cost. They prevent harmful outcomes that might arise from an exclusive focus on the north star. For a social media app focused on user engagement as its north star, a guardrail metric might be monitoring instances of inappropriate content or user reports to ensure user safety. For a ride sharing app like Uber or Lyft, a guardrail metric might be the number of support tickets created per week.

Common Metrics

Acquisition	Activation	Engagement & Retention	Revenue	Referrals
Traffic Source	User Registration Rate	Daily / Monthly Active Users (DAU/MAU)	Average Revenue Per User (ARPU)	Net Promoter Score (NPS)
Cost Per Acquisition (CPA)	Time to First Key Action (Aha Moment)	User Churn Rate	Customer Lifetime Value (LTV)	Referral Rate
Conversion Rate		Session Length		Virality Coefficient
	Onboarding Completion Rate			
Click-Through Rate (CTR)		Frequency of Use	Free to Paid Conversion Rate	Social Share Count
	Account Setup Completion Rate			
Social Media Engagement / Reach		Product Stickiness	Monthly Recurring Revenue (MRR)	Invite Acceptance Rate
App Install Rate		Feature Adoption		

Examples of common metrics, adapted from a LinkedIn post by Siddharth Arora[154].

How Do Other Metrics Work with North Star?

There are four focus areas of a metric — **value, breadth, depth, and trend** — interact with north star, signpost, and guardrail metrics in meaningful ways, providing a comprehensive understanding of a product's performance.

At its core, a north star metric reflects the fundamental value a product delivers to its users. This metric aligns directly with the value focus area, as it encapsulates the ultimate benefit users derive from the product. For instance, for Tinder, the number of successful matches made would be a perfect north star metric, mirroring the inherent purpose and value proposition of the app.

Signpost metrics, on the other hand, provide a more detailed picture of the product performance, tying into the breadth and depth focus areas. Breadth metrics give us an understanding of how many users are finding value in the product, such as the count of Daily Active Users or Weekly Active Users. These metrics often serve as signposts, indicating whether the product is reaching a wide user base and where there might be room for growth. Depth metrics, like the number of sessions per user or retention rate, reveal how deeply and frequently users engage with the product. These too can be crucial signpost metrics, shedding light on user behavior and loyalty, and suggesting areas for improving user experience and retention.

Trend metrics, which track changes over time, could serve as either signpost or guardrail metrics, depending on their focus. As signposts, trend metrics can help predict future performance based on past and current trends, for example, by tracking growth rate of unique visitors month-over-month. On the other hand, as guardrails, they could help identify and prevent potential issues. If the retention rate across different cohorts is trending downwards, this guardrail metric would signal a need to investigate and address possible causes.

In essence, value, breadth, depth, and trend focus areas provide different lenses to view product performance, and each lens aligns with north star, signpost, or guardrail metrics in unique ways. Together, they form a comprehensive framework for measuring success and driving growth for any product.

Input and Output Metrics

Input and output metrics are invaluable tools for gauging product performance. They're closely related to your focus areas, as well as your north star, signpost, and guardrail metrics.

Input metrics, which measure what you put into a system, often align closely with breadth and depth metrics. For example, consider "customer service response time" as an input metric. A quick response time is aimed at deepening customer engagement and satisfaction (depth). Such input metrics can act as signpost metrics, giving you early hints on whether your strategies are successfully engaging users.

Output metrics, conversely, measure the results your efforts produce. These metrics usually align with value and trend focus areas. They often directly result from your input metrics and capture the value your product offers to users. For example, customer satisfaction scores (value) or the rate of customer retention month-over-month (trend) could be output metrics. Your North Star metric is likely an output metric, as it epitomizes the highest value your product delivers to users.

Guardrail metrics can be either input or output metrics. These metrics act as a safety net, making sure your pursuit of North Star and Signpost metrics doesn't lead to negative consequences. If your input metric is "customer service response time," a guardrail metric could be tracking the quality of those customer service interactions to ensure speed doesn't compromise service quality.

Input and output metrics offer a way to understand the cause-and-effect relationship within your product ecosystem. Used alongside focus areas and key metrics like the North Star, Signposts, and Guardrails, they offer a holistic view of your product's performance,

Unique Generative AI Considerations

When it comes to measuring success in generative AI products, the rule is mostly the same as with traditional AI or software offerings: focus on delivering real customer value and ensure healthy feature adoption and retention. However, there are a few unique considerations to keep in mind when setting your North Star, Signpost, and Guardrail metrics. It's crucial to gauge the actual impact of generative AI features to differentiate between meaningful progress and mere trends or vanity additions.

1. Complexity of Output: Traditional metrics like user engagement or retention might not fully capture the performance of generative AI. The output of these AI systems can be quite complex, making it challenging to measure success. For example, for a language model like GPT-4, the north star metric could be the percentage of completed tasks where the user found the AI's output useful. The definition of "useful", however, can be subjective and may require further elaboration. Signpost metrics could measure individual aspects of the output that contribute to the overall value, while guardrail metrics must ensure that the complexity doesn't lead to negative outcomes, such as % of users reporting confusion or issues.

2. Quality, Novelty, and Diversity: As discussed in the previous section, success of generative AI products often hinges on the quality, novelty, and diversity of generated content. These dimensions can form a part of your signpost metrics. For instance, a generative music app might have a north star metric related to the number of users regularly using its compositions, but signpost metrics could include user ratings of song quality, perceived uniqueness of each composition, or the variety of musical styles generated.

3. Interpretability and Fairness: As a core content of responsible AI, it's important to consider how well users understand the AI's output (interpretability) and perceive the output is fair and unbiased. For example, for an AI that generates resume feedback, the north star metric might be the number of users who improve their resumes based on the AI's suggestions, whereas signpost metrics could include user surveys on the clarity of feedback and guardrail metrics should monitor any reports of perceived bias or unfairness from AI suggestions.

4. Continuous Evaluation: Given the evolving nature of generative AI models, north star and signpost metrics should reflect this dynamic performance over time, requiring regular measurement and updates. Guardrail metrics, in this context, would ensure that changes in the model's performance don't lead to undesirable consequences.

5. User Interaction: Generative AI products often interact directly with users, making the user experience a vital part of the evaluation. User feedback and sentiment could serve as additional signpost and guardrail metrics, providing insights into how users perceive the value generated by the AI.

Metrics are the compass for product management, guiding teams toward their goals. Understanding various metrics—focused on value, breadth, depth, and trend—enables better monitoring and success. For generative AI products, once and again, it's crucial to measure real impact to separate meaningful progress from mere hypes. Balancing input and output metrics shapes informed strategies that convert efforts into results. Remember, what gets measured gets managed, especially true in product management.

Case in Point: The Fall of Kite

Kite, an AI company, was first out of the gate with a coding assistant back in 2020—18 months before GitHub Copilot came onto the scene. With machine learning at its core, Kite's platform was like a valuable toolkit for developers, offering real-time code suggestions, error detection, and documentation. Its line-of-code completions were honed on over 20 million open-source files, and its code snippet recommendations were downright clever. Within a short span, Kite amassed a sizable user base of 500,000 monthly-active developers—all with almost zero marketing spend.

So far, so good. But here comes the snag.

The CEO revealed in a blog post that Kite failed to monetize its user base. Individual developers weren't willing to pay for the tool, and engineering managers, who held the purse strings, were lukewarm about Kite's offering. Despite making developers 18% faster, this wasn't deemed compelling enough to warrant a purchase.

Then the team attempted to pivot, exploring the avenue of code search, but by that point, after seven grueling years, the team was burnt out. Kite sought a "soft landing" instead of forging ahead with the new direction.

What is the REAL problem here?

At the heart of this downfall was Kite's nebulous north star. GitHub Copilot, by contrast, was crystal clear: "Improving Developer Productivity" was its north star. Now, let's make it more specific: "Maximizing Developer Output per Hour." While not a verified metric, it's plausible that GitHub Copilot uses something similar to this as a more refined north star to guide its decisions.

Consider the economics: A software developer earning $200,000 per year would find immense value in a 18% productivity boost—equivalent to $40,000 annually. Wouldn't such a developer gladly pay a $100 annual subscription for that advantage?

Had Kite aligned its priorities with a well-defined north star like "Maximizing Developer Output per Hour," the story might have been quite different. Their focus would have shifted, guiding them toward creating features and experiences that developers—not just managers—would find indispensable enough to pay for.

So, the cautionary tale of Kite serves as a potent reminder: A well-defined north star isn't just a lofty ideal; it's a critical compass that could mean the difference between thriving and merely surviving[155].

Note: This case study is an interpretative analysis and may not encompass all factors of Kite's journey. We welcome additional insights to enrich this case study.

2.32 - Why Do Promising Products Fail at Go-to-Market (GTM)?

Picture this: A bustling tech startup in the heart of San Francisco. A team of bright engineers and product managers have just put the finishing touches on a groundbreaking AI product. It promises to revolutionize an industry, to make the impossible possible. The conference rooms buzz with discussions about potential revenues, market reach, and competitive edge – and yet, six months later, the company vanishes from the limelight, the revolutionary product barely making a ripple in the market. What happened?

Welcome to the perplexing world of Go-to-Market (GTM) strategies, where even the most promising products can falter and fade away if not maneuvered correctly into the marketplace. The term GTM often conjures images of sales teams hustling and marketing mavens crafting compelling narratives, it's crucial to recognize this as a symphony rather than a solo act. While sales and marketing are the indispensable instruments that set the rhythm and tone, product leaders serve as the conductors, orchestrating the various elements to create an integrated experience. Together, these three pillars are collectively responsible for ensuring that the product not only solves a problem but also finds its way into the hands of those who need it most.

Up to 90% of startups fail, and AI startups are no exception[156]. A significant number of them falter precisely at the go-to-market (GTM) stage. So, what are the common root causes of these failures?

Common GTM Challenges

Shallow Market Research: One of the biggest mistakes is not conducting thorough market research before launching their product. This cursory approach leaves them without the deep understanding needed to truly grasp market demands, customer pain points, and the competitive landscape. Without this nuanced insight, crafting a compelling value proposition becomes challenging. Moreover, this lack of depth can be compounded by a common blind spot: falling too deeply in love with your product. This emotional attachment can cloud judgment and make teams ignore signs that their solution may not adequately solve a customer's problem. In turn, the product risks becoming a solution in search of a problem, rather than a meaningful answer to a real, pressing issue. Investing the necessary time and resources into deep market research helps identify true customer's pain points and informs the development of a unique selling proposition.

Inadequate Customer Segmentation: Another mistake often made in GTM strategies is failing to properly segment the target audience. Without a clear understanding of customers' diverse needs, preferences, and purchasing behaviors, product teams may fail to tailor marketing messages and deliver a personalized customer experience. Segmenting customers based on demographics, psychographics, and behavior can lead to more targeted marketing campaigns and improved customer engagement.

Weak Value Proposition: An underwhelming or value value proposition can be the undoing of even the most meticulously crafted go-to-market strategy. This linchpin statement needs to do more than just exist; it must clearly articulate how the product alleviates customer pain points, offers unique advantages, and differentiates from the competition. When this vital message falls short, the outcome is often a sea of perplexed prospects and squandered opportunities – but clarity is only half the battle. A compelling value proposition must also be communicated in a manner that resonates with your target audience. Whether through captivating storytelling, persuasive statistics, or concrete-benefits-focused framing, the style in which you convey your proposition is as critical as the substance itself.

Ineffective Pricing Strategies: Pricing is critical to any GTM strategy. However, many companies make the mistake of setting prices without considering market dynamics and customer perceptions. Pricing should align with the value delivered and closely consider competitor pricing, customer willingness to pay, and product differentiation. Failing to establish the right pricing strategy can lead to lost customers or leaving money on the table.

Insufficient Marketing and Sales Alignment: Proper alignment between marketing and sales functions is essential for a successful GTM strategy. However, many organizations struggle with this alignment, resulting in missed opportunities and decreased revenue. It's not just the responsibility of sales and marketing to bridge this gap; product leaders too must step into the fray. By collaborating closely with these teams and contributing product and customer insights, strategies, and milestones, product leaders can help forge a unified approach. This three-way partnership cultivates consistent messaging, refines the customer journey, and ultimately propels conversion rates to new heights.

Lack of Customer Onboarding and Retention Strategies: Acquiring new customers is only part of the equation; retaining and nurturing those customers is equally important. However, companies often neglect this aspect of their GTM strategy, resulting in high churn rates. Implementing effective customer onboarding processes, providing ongoing support, and nurturing customer relationships is crucial for driving long-term growth and maximizing customer lifetime value.

Fail to Continuously Iterate and Adapt: In the fast-paced world of SaaS in particular, a static GTM strategy is a recipe for obsolescence. The marketplace is always evolving, marked by fluctuating customer needs and preferences. SaaS companies must regularly review and adjust their GTM strategies, adopting an agile approach informed by ongoing feedback. The rule of thumb is straightforward: either continuously improve or risk becoming obsolete. This is exacerbated in the ever-changing realm of AI, where new algorithms and technologies can quickly render your product outdated, compelling swift iterations or even drastic shifts in direction.

Underestimate Public Awareness and Educational Gaps of AI: A major GTM challenge in launching AI products lies in the dual hurdles of varying AI understanding and the ability to clearly explain AI functionalities. Product teams immersed in tech circles often wrongly assume a universal comfort with AI, neglecting the diverse levels of public perception and trust. This problem is compounded by the inherent complexity of AI technologies, which necessitates a broad educational initiative to even describe what the product does. This creates not only a slow time-to-market but also amplifies PR risks, especially for established companies with an existing user base. Failing to address these intertwined issues can result in a disconnect between the product and its potential users, hampering both adoption and trust.

How to Do GTM Right?

Below is a list of principles/best practices for GTM:

Establish a well-defined GTM funnel. Securing your first three to five paying customers through an MVP signifies some level of product-market fit. However, relying on sporadic, founder-led sales won't bring a consistent inflow of customers. To transition from ad-hoc efforts to a sustainable acquisition strategy, you'll need a well-defined GTM funnel. As Keith Messick from LaunchDarkly points out, once a company scales beyond $10M, the sales process transforms into a quantifiable system. The first step is to set up comprehensive funnel reporting. Rooted in demand generation, this reporting structure classifies leads—often within a CRM—into categories like Marketing Qualified Leads (MQL) and Sales Qualified Leads (SQL), while also defining opportunity stages such as 'closed/won' and new Annual Recurring Revenue (ARR). Collecting and organizing this data is vital. It not only helps you gauge the health of your funnel at various stages but also informs future investment decisions. The sooner you formalize this data collection, the smoother your transition to a sales-driven approach will be.

GTM success is everybody's job. In today's fast-changing market, a robust GTM strategy hinges on seamless collaboration between sales, marketing, and product management. Though these functions focus on different aspects—sales on lead conversion, marketing on brand awareness and lead generation, and product on user experience—their goals are interconnected. You have two options for organizing your team for this collaborative effort. One approach is to form a dedicated growth team with roles like growth PM, engineers, and designers, focusing them on specific Product-Led Growth (PLG) metrics such as signups and onboarding. Another is to create a 'tiger team' made up of experts from all functions for tackling big projects like launching a freemium product or setting up a Product-Qualified Leads (PQL) system (more on this later).

Regardless of the team structure you choose, the key to harmonious GTM lies in tracking funnel metrics uniformly across all departments, setting collective goals, and putting in place transparent processes for transitioning leads through the funnel. Criteria or leadership approvals for moving a lead from one stage to another ensures everyone is aligned. This holistic approach reduces the risk of bottlenecks in any single function and allows for swift adaptations to market changes, making each department a stakeholder in the company's overall success.

Laser focuses on educating customers and delivering value. In any GTM strategy, the value of customer success is often underrated—especially in industries like generative AI and AI more broadly. While flashy demos can catch the eye and generate buzz, delivering real value involves navigating intricate AI infrastructures, closing feature gaps, and achieving end-to-end business impact. Here's the kicker: none of this can be accomplished in silos. Ensuring that customers truly understand and gain tangible benefits from complex AI technologies requires an orchestrated effort across all departments.

In addition, to set your customers up for success, it's essential to educate them about the AI technology they'll be using. Manage expectations carefully by breaking down what the technology can and cannot achieve, thereby ensuring a smoother adoption process and a longer-term relationship. By aligning sales, marketing, product, and customer success around customer education and managing expectations, companies can not only prevent any single function from becoming a bottleneck but also adapt more agilely to market shifts. This unified approach makes each department a stakeholder in the customer's success, transforming customer value from a tagline into an operational reality.

Make community part of your GTM strategy. In today's hyper-connected world, integrating community-building into your GTM strategy is not just an afterthought; it's a growth multiplier. When you invite your customers to be an active part of your journey—whether it's providing feedback, previewing new features, or simply giving them a space to talk and connect— you're doing more than just customer research. You're fostering trust, and this trust serves as a robust engine for product development, customer acquisition, and retention.

Take Notion, for example. Their community forums and tutorials are not just customer service channels but innovation hubs. Users often conceptualize new features and solve each other's problems. This not only informs Notion's product roadmap but also creates a loyal customer base that's invested in the platform's success[158].

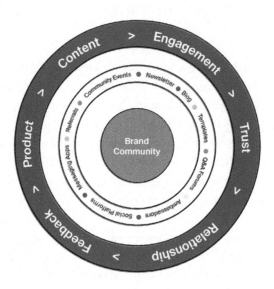

The community-led growth flywheel. Image from No Good[157].

The community-led approach is particularly compelling for generative AI products, as it cultivates an ecosystem where users not only provide data but also generate insights and custom solutions. Imagine having a community that's part of your iterative model training, providing real-world use-cases that make your AI smarter with each interaction. Midjourney is a terrific example; it uses Discord to turn its 14-million strong community into a dynamic extension of its GTM strategy, enriching its AI models with real-world interactions. This mutual value creation amplifies product-led growth and turns your community into a dynamic extension of your GTM strategy. Incorporating a community into your GTM strategy, you're not just building a product; you're creating a living ecosystem that thrives on shared value and mutual growth[159].

Know when the demand shifts from inbound to outbound and adapt accordingly. During the early days of a company, particularly those experiencing strong product-led growth and bottom-up sales, the focus is often on creating an outstanding product that naturally generates a high volume of inbound requests. In this "inbound-limited" phase, the challenge usually isn't acquiring demand; it's effectively managing and converting that demand into revenue.

As the company scales and revenue increases, inbound growth often plateaus, transitioning the business into an "outbound-limited" stage. At this point, it becomes crucial to establish a scalable demand generation engine. This involves maximizing demand through existing channels and exploring new avenues like SEO, events, webinars, and paid search. Additionally, implementing lead scoring and conversion tracking becomes essential to efficiently manage this new, more complex stage of demand generation.

Refine your pricing strategy to align with customer value and opt for simplicity. While companies can get away without a formal pricing and packaging strategy at the initial stage, it can become a significant roadblock to growth as pipeline velocity picks up. Ad hoc pricing can slow down deal momentum and confuse reps trying to execute quickly on a day-to-day basis.

As your company grows, transition from ad hoc pricing to a structured model when you have sufficient data to understand your customer segments and growth drivers. Implement simple and clear pricing tiers or packages that reflect how customers derive value from your offering, such as per user, per usage, or value-based pricing. Equip your sales reps with streamlined discount guidelines to expedite deal closures. Take Notion AI as an example. They've effectively communicated the value of their AI features, leading them to adopt a unified usage-based pricing model for all customers. This clear and simple pricing strategy has fueled their rapid growth, without the need for complex enterprise sales negotiations at the outset.

Spotlight: Pricing Challenges for Generative AI Products

Unlike traditional software products that can enjoy near 100% gross margins, AI products, as exemplified by GitHub's Copilot, can be loss-making endeavors. According to the WSJ article, while Microsoft charges $10 monthly for GitHub Copilot, the company loses an average of $20 per user each month, with heavy users costing up to $80 a month. The real-world compute, electricity, and API costs challenge the conventional "low cost, high margin" model of the software industry[160].

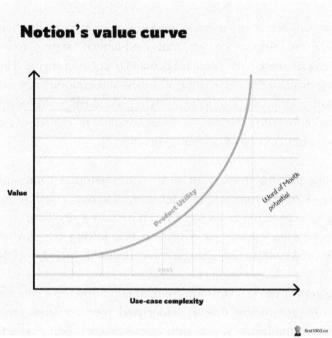

This is a great chart illustrating how traditional software products can continue to move users along the value curve with new use cases and product capabilities without incurring additional cost. Image from Ali Abouelatta and First 1000[161].

An immediate solution is token-based pricing, letting companies pass on fluctuating costs to consumers. While token-based pricing might make sense at smaller scales or during the early stages, it lacks the predictability that enterprise clients prefer. In contrast, traditional Software-as-a-Service (SaaS) models provide this predictability through flat-fee subscriptions, something not easily replicated in AI services with variable costs.

Incumbents with deeper pockets and a larger customer base bet on falling AI cost. They can afford to underprice AI products to improve their models through data volume. This makes it hard for startups to compete unless they adopt counter-positioning monetization strategies. Value-based pricing appears to be the early winner. As Benchmark partner, Sarah Tevel, suggests, "sell to work, not the software[162]." This means pricing based on the work output, like legal agreements, instead of per seat. This opens up new markets and allows for better perceived value.

Claire Vo, the CPO at Color and former CPO at Optimizely, in her interview with Reforge's Unsolicited Feedback podcast with Brian Balfour and Fareed Mosavat, predicted a future where automation decreases the perceived value of human expertise. In such a landscape, traditional productivity-based pricing could face a "death spiral" as department heads will increasingly weigh budget allocations between humans and AI[163]. We will explore how to differentiate and build moats for generative AI products in section 2.4.

These insights highlight the multifaceted challenges of pricing AI products and its evolving dilemma that the industry has yet to fully solve.

2.33 - Choosing the Right Growth Strategy: Product-Led Growth (PLG), Marketing-Led Grow (MLG), or Sales-Led Growth (SLG)?

The debate over the optimal growth strategy—Product-Led Growth (PLG), Marketing-Led Growth (MLG), or Sales-Led Growth (SLG)—is a staple in go-to-market discussions and there are many schools of thought. One school of thought, as per SaaS expert David Sacks, is an approach rooted in founder competencies[164].

David Sacks ✓ 🔳
@DavidSacks

Founder/Go-to-Market Fit:

If primary GTM is bottom-up, founder needs to be great at product.

If primary GTM is inbound, founder needs to be great at marketing.

If primary GTM is outbound, founder needs to be great at sales.

5:26 PM · May 11, 2021

307 Reposts   **49** Quotes   **2,029** Likes   **657** Bookmarks

Another primary school of thought is based on what it takes to build a $100 million company based on the Annual Revenue Per User (ARPU) - a common benchmark among many Silicon Valley venture capitalists when evaluating startups. This idea is simplified as the "hunting method," which groups customers into five animal types: flies, mice, rabbits, deer, and elephants. Each animal represents a different path to reach $100 million in yearly sales.

Flies ($10 per user x 10 million customers) are the smallest, requiring millions of users for significant revenue. The primary growth lever here is *virality*. Companies like Instagram excel by creating features that users eagerly share, driving new sign-ups.

Mice ($100 per user x 1 million customers) require fewer users but higher ARPU. The primary lever for this group is subscription, often led by product quality. Spotify, for instance, offers a premium subscription model that entices users to pay for an ad-free experience.

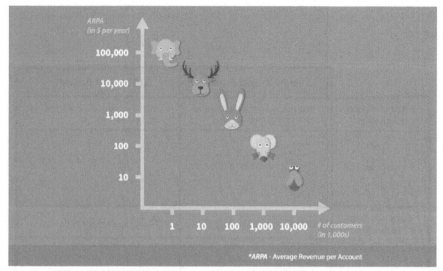

Infographics from Jeff Desjardins and Visual Capitalist[165].

Rabbits ($1k per user x 100k customers) are typically small businesses, and the main growth lever is inbound marketing. High-quality content and strong SEO strategies help attract these types of customers who require a product that solves their specific needs efficiently. In the early days, HubSpot offered dozens of free marketing courses designed to build good relationships with marketers, who in turn trust HubSpot with their businesses.

Deer ($10k per user x 10k customers) are medium-sized businesses requiring a more nuanced sales approach. Your customers are getting big enough that they will demand some human contact, yet not worth sending a huge sales force for company visits. A common intermediate tactic is inside sales or remote sales, where customers do get to speak with a sales rep, but mostly only over the phone.

Elephants ($100k per user x 1k customers) are large enterprises, and the primary growth lever is enterprise sales. These are complex, long-term sales cycles requiring a specialized sales team. Companies like Salesforce shine here, offering robust solutions that can be deeply customized to meet enterprise-level needs.

The choice between PLG, MLG, or SLG depends on which 'animal' you're targeting[165].

	Product-Led Funnel	Sales-Led Funnel
Initial customer segments	Prosumers, SMB, or small teams within companies	Mid-size companies and enterprises
Funnel metrics	• # of free signups • # of activated users/teams • Free-to-paid conversion rate • # of PQLs • # of SQLs • ARR	• # of leads • # of MQLs • # of SQLs • ARR
Early success indicator	Product usage	Marketing engagement
Growth methods	• Primarily rely on product experience and automated lifecycle marketing (e.g. emails, in-app) to drive usage • Reserve sales touch for selected high-value customers	• Primarily rely on marketing campaigns to nurture leads • Sales broadly involved to qualify leads and close deals
Potential teams involved	• Growth Marketing • Lifecycle Marketing • Product Growth • Sales (online, SMB)	• Demand Generation • ABM Marketing • Sales (Enterprise)

Major differences between PLG and SLG. Image from Hila Qu, Lenny's Newsletter[166].

Similarly, growth expert Hila Qu argues the choice between PLG and SLG should hinge on your target customer segment and product complexity. For prosumers and small businesses, PLG is often the most efficient growth strategy, while complex enterprise solutions may initially favor SLG. Visualizing a PLG funnel will help you understand the customer journey and what adjustments your team needs to make.

Growth expert, and ex-Head of Growth at Amplitude, Miro and SurveyMonkey, Elena Verna introduces the third school of thought, which reframes the discussion around growth not as a simple choice between PLG, MLG, or SLG.

Instead, she likens it to crafting a 3-course meal[167] consisting of acquisition, retention, and monetization. In this culinary metaphor, each growth model— PLG, SLG, MLG—is like a basket of ingredients from which you can choose to make a comprehensive growth menu. The acquisition serves as the appetizer, vital but leading to the main course, which is retention. Monetization is your dessert, a rewarding finish only achievable by nailing the first two courses. Beginners might start by sticking to one basket, but the endgame is to be a master 'chef,' adept at mixing ingredients from all models to serve up a unique and robust growth strategy[167].

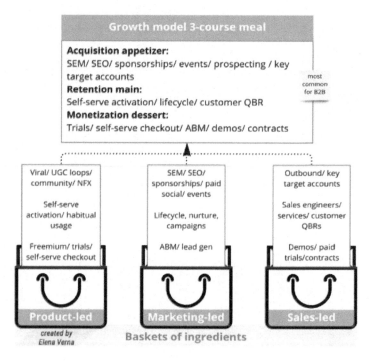

Image from Elena Verna[167]

Last not least, growth expert Leah Tharin went as far as mapping out the mental model in a complex flow chart[168]:

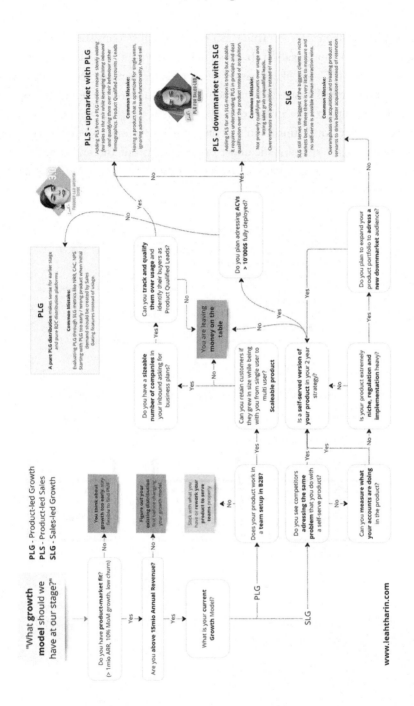

While many companies use a blend of PLG, MLG, and SLG strategies, the rise of generative AI brings unique considerations. Traditional software offers several advantages, such as high gross margins and defensible intellectual property. However, AI's data-centric nature introduces extra costs:

1. **Infrastructure:** Beyond regular backend and system integration costs, AI incurs expenses for regular training and ongoing inference, especially in the Language Learning Model (LLM) space.

2. **Human Oversight:** The Machine Learning Ops (MLOps) lifecycle demands continuous human interaction to ingest data, monitor performance, retrain models, and deploy updates.

Given these factors, AI business growth will likely require a hybrid of PLG and either MLG or SLG, leaning toward the latter two due to the service-oriented nature of AI offerings. However, to keep AI as attractive as software, it's crucial to automate or "productize" service elements.

For generative AI businesses to scale effectively, the core growth strategy should focus on PLG, supplemented by elements of MLG and SLG.

What Is PLG and How to Get Started?

What Is PLG?
Elena Verna defines PLG as "Growth is your ability to *predictably*, *sustainably*, and *competitively* answer the question of how you acquire, retain, and monetize customers."

How to Start a Successful PLG Motion?
According to Amplitude, a leading Product Analytics & Event Tracking Platform, to kick off a PLG strategy, your product must be at the heart of the entire customer journey. This requires an in-depth understanding of various touchpoints, such as:
- Your most effective acquisition channels and the reasons they excel.
- Points of friction during onboarding.

- The pathway to the customer's "Aha" moment.
- Instances where user engagement drops off before activation.
- Features that are crucial for retaining users.
- The transition journey from a free user to a paying customer.

Based on this understanding, prioritize your primary growth constraint, be it acquisition, retention, or monetization, and make that your starting point. This targeted approach ensures maximum ROI. While *acquisition* uses the product to generate leads and relies on viral loops and content loops, *retention* hinges on a successful activation, guiding users to the "aha moment" as quickly as possible through a multi-channel approach that includes onboarding, emails, and human assistance. *Monetization* aims to turn free users into paying customers either through self-service or by identifying Product-Qualified Leads (PQLs) for the sales team to follow up on. Prioritize these aspects based on your most immediate needs for fast, impactful results.

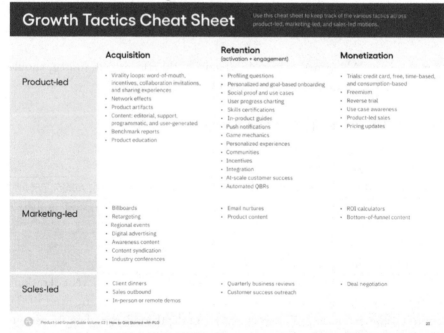

Source: Amplitude's "Product-Led Growth Guide Vol. 2: How to Get Started with PLG"

Above is a summary of various growth tactics companies can employ across product, marketing, and sales from Amplitude. We've covered some of the product-led tactics in the "Designing based on engagement states" section 2.24 of the book.

Besides identifying the right growth lever, for PLG to succeed, you need three core components:

1. **Data**: A culture that embraces data-driven decision-making, ongoing experimentation, a growth mindset, and radical honesty. Are you prepared to face uncomfortable truths if your product underperforms? In growth, failure is a critical path.

2. **Technology**: An advanced growth tech stack that facilitates user access, identifies onboarding friction, nurtures relationships via personalized communications, and leverages product usage data for lead scoring.

3. **Skills and Capabilities**: Teams composed of growth "ninjas" proficient in understanding product nuances, user psychology, copywriting, customer modeling, data analytics, and research. These individuals should also be endlessly curious and capable of thinking systematically about how all growth levers interact.

Introducing Product-led Sales (PLS)

Adopting PLG doesn't necessitate sidelining your sales or marketing team; it calls for a fundamental paradigm shift. Unlike traditional "top-down" sales models focused on landing a few big enterprise clients, PLG is a "land-and-expand" strategy. It starts by offering free or low-cost entry plans to build a broad user base. Over time, the most engaged and promising of these free users are converted into paying customers. This doesn't cannibalize traditional sales revenues; instead, it enhances the quality of the sales pipeline through genuine product engagement metrics. If you're looking for a harmonious blend, consider implementing a Product-led Sales (PLS) approach, which capitalizes on PLG efficiencies while still engaging the prowess of a sales team[170].

Growth expert Elena Verna illustrated the Product-Led Sales journey brilliantly in this diagram:

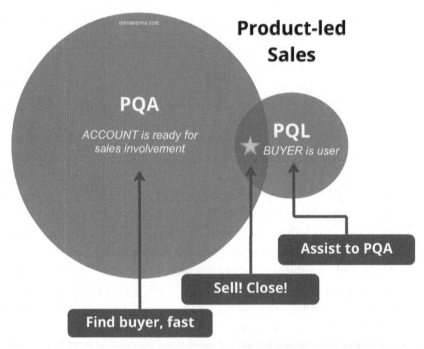

Image from Elena Verna[171].

While Product Qualified Accounts (PQA) pinpoints when an account is primed for sales engagement by tracking its journey from sign-up through regular engagement, Product Qualified Leads (PQL) helps identify if the person using the product is the right buyer. When both PQA and PQL are in place, closing a deal is generally swift and straightforward.

However, it's essential to approach different scenarios strategically. For instance, when you have PQA but not PQL, targeted Account-Based Marketing (ABM) can help you find the right buyer and connect them to the product. When PQL is present but PQA is lacking, don't overlook potential enterprise buyers; guide them to become engaged users before making the sales pitch.

Learn more the PQAs and PQLs at https://www.endgame.io/blog/elena-verna-pqa-pql-guide. Find out more about when PLG companies are ready for product-led sales here: https://www.endgame.io/blog/elena-verna-ready-for-pls.

Case in Point: PLG in Action at Amplitude

Amplitude has optimized its PLG strategy by focusing on the activation stage of the customer journey. Key tactics include simplifying the onboarding experience, leading to a 60% increase in signup-to-account creation conversion. Amplitude also launched a Sales Assist Program and customized onboarding based on user goals. These efforts were complemented by goal-driven email nurture flows and community resources like Amplitude Academy and an online forum. These initiatives made the setup process seamless, with one-click integrations for data sources adding further momentum.

Metrics form the backbone of Amplitude's PLG strategy. Led by Head of Growth Marketing, Franciska Dethlefsen, the company tracks key performance indicators like new relevant signups for acquisition and activated accounts for retention. The "set up moment" is identified as a successfully connected data source and chart creation, while the "Aha" moment is marked by saving or sharing a chart. Monetization is evaluated by the free-to-paid conversion rate over a rolling 28-day window, offering a comprehensive view of how Amplitude's PLG strategy drives sustainable growth.

CATEGORY	METRIC
Acquisition	Number of new signups and/or qualified leads
	Customer acquisition cost (CAC)
	Payback period
Activation	Activation rate
	Time to activate
	Free-to-paid conversions
Engagement	Monthly, weekly, and/or daily active users (MAU, WAU, DAU)
	Stickiness (DAU, MAU)
	Feature usage
Retention	Retention rate
	Churn rate
	Customer lifetime value (CLV)
Monetization	Net revenue retention (NRR)
	Monthly recurring revenue (MRR)
	Average revenue per user (APRU)

Source: Amplitude's "Product-Led Growth Guide Vol. 2: How to Get Started with PLG"[172].

When NOT to Use PLG?

Product leader Lindsey Liu argued that while Product-Led Growth (PLG) is a compelling strategy, it's not a one-size-fits-all solution. Here are some scenarios where PLG may not be your best bet:

1. **Premature Growth Before Achieving PMF:** If your product hasn't yet achieved PMF; don't waste time fueling the leaky bucket.

2. **Complex Onboarding:** If your product isn't self-serve, demands technical implementation or requires human intervention for activation,

PLG is probably not the right path. New users should be able to easily try the product without a massive investment of time or effort.

3. **Data Quality and Scope**: PLG thrives on high-quality data for experimentation. If your product mainly attracts small businesses but aims for enterprise-level clients, the data will be skewed. This creates a mismatch between your product's positioning and your ultimate target.

4. **Top-Down Decision Making**: In cases where value is derived only when an entire organization adopts the product, like intranet solutions or enterprise HR software, top-down buying decisions are inevitable. PLG isn't designed for such long sales cycles.

5. **Low User Engagement and Perceived Value:** If your users only interact with the product sporadically, they're unlikely to become champions who advocate for its broader adoption. For example, a team using a scheduling tool just for occasional lunches won't push for company-wide use. Similarly, if the value of your product isn't immediately obvious—like a cybersecurity app that operates in the background—users are less likely to fully embrace it. Both scenarios limit the effectiveness of a PLG strategy.

6. **Regulatory Constraints and Special Requirements:** Some industries like healthcare and government are highly regulated and may not support a bottoms-up expansion strategy.

Remember, the effectiveness of PLG is contingent on several factors, including user behavior, data quality, and organizational structures. Make sure to weigh these elements carefully before diving into a PLG approach[173].

2.34 -Putting It All Together: The PLG Iceberg & Canva's Growth Story

Introduction

The best way to bring all the PLG concepts together is by looking at the PLG Iceberg. The concept was brilliantly conceived by Jaryd Hermann in his newsletter "How They Grow." Hermann's framework serves as a robust lens through which we can examine how companies like Canva have leveraged multi-layered strategies for exceptional growth.

Now, let's embark on an exploration of Canva's PLG iceberg, where each layer—from L1 to L8—serves as a blueprint for product-driven success.

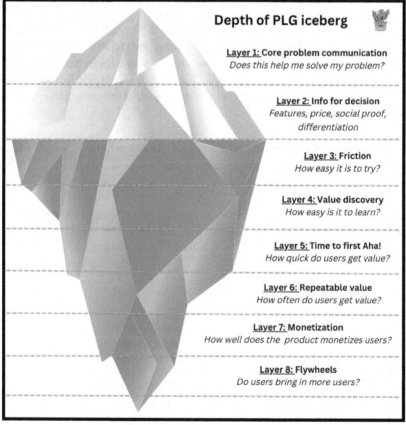

Image from How They Grow[174].

Founding & Initial Growth: Identifying the Tip of the Iceberg

Canva was born from the idea that design should be democratized. Back in 2013, the founding team led by Melanie Perkins saw an unmet need—small businesses were flocking to Facebook, but there was a glaring gap between professional design tools like Adobe and simplistic templates. They created Canva as a tool for anyone to create compelling visuals for social media, precisely when the demand was peaking.

Crossing the Chasm: Targeting High-Expectation Customer

The early days were all about "crossing the chasm" from early adopters to a mainstream audience. Canva meticulously selected their 'High Expectation Customer (HXC)'—social media marketers and bloggers who lacked design expertise but needed professional visuals.

As Jaryd Hermann put it, the key to crossing the chasm for Canva is positioning and securing a "beachhead" in a mainstream market. Simply put, your beachhead is the smallest customer segment you can target that serves as a point of attack into the main market you want to break into. This niche has a set of characteristics that make it an ideal place to focus your initial effort.

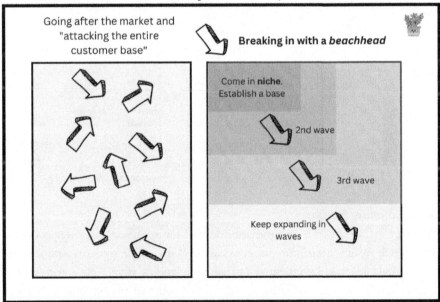

Image from How They Grow[175].

This focus paid off; Canva became the go-to tool for designing Facebook graphics, which not only addressed a pain point but also became a distribution channel in itself. This was brilliant as it allows Canva to tap into viral growth through easy sharing and word-of-mouth, turning a narrow focus on social media design into a gateway for broader user adoption.

L1+ L2: Communicating Value

When the chasm was crossed and Canva was ready to shine at the iceberg's tip, Layers 1 and 2 of the PLG Iceberg are about effectively communicating value to users and helping them decide to try the product. When someone googles a design-related query, like "how to make a logo," they're directed to a Canva landing page that is laser-focused on solving that specific problem. Each of these optimized landing pages touches on particular value points tailored to the user's needs, making it abundantly clear why Canva is the go-to solution. This effective communication of value is a cornerstone of Canva's PLG strategy, setting the stage for deeper engagement in the layers that follow.

Laser focus on solving specific problems and showcasing specific use cases

L3: Removing friction

In Layer 3, Canva excels in removing friction during the onboarding process to effortlessly usher users into its ecosystem. With single-sign on options and contextual cues based on the user's originating landing page, Canva immediately suggests relevant templates to speed up the journey to the 'Aha!' moment. Within a minute, users can transition from a landing page to a working design canvas. Onboarding isn't just confined to the platform; Canva employs timely email nudges for users who haven't taken action, followed by weekly newsletters featuring new templates, design tips, and upsell opportunities for Canva Pro. The focus is on education rather than direct selling, aligning the user closer to repeated value and solidifying Canva's standing in the third layer of the PLG Iceberg.

L4 + L5: Value Discovery and 'Aha!' Moments

In Layers 4 and 5, Canva hits the sweet spot between value discovery and crafting 'Aha!' moments. Unlike intimidating platforms like Adobe Illustrator, Canva's easy-to-use interface and "great defaults" dramatically shorten the learning curve. This user-friendliness, combined with an expansive template library, enables users to quickly create designs they're proud of—often in just minutes rather than hours. This speed to value not only captures immediate user satisfaction but also creates a shareable 'Aha!' moment, amplifying word-of-mouth and bolstering Canva's strength in these crucial layers of the PLG Iceberg.

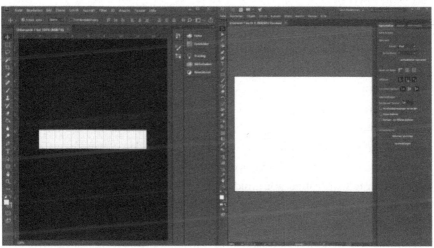

It can be intimidating to learn all the available tools available in Adobe

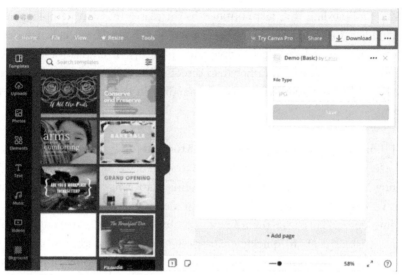

It's much easier to get inspired and start with well-designed templates

L6: Repeatable Value

In Layer 6, Canva demonstrates the art of sustaining value, a critical component for long-term business viability. Unlike models that limit user engagement through paywalls or restricted usage, Canva's freemium model offers unrestricted access, facilitating habit formation. The platform achieves stickiness through three strategic moves: 1) covering a diverse range of design needs, from logos to social media posts, all integrated into one platform; 2) vertically integrating the design value chain to offer a one-stop-shop experience, negating the need for multiple tools; and 3) continually evolving its offerings in response to users' changing behaviors and needs, such as the adjustments made for the remote work landscape amid the COVID-19 pandemic. Together, these elements create a 'sticky ecosystem' that not only attracts users but keeps them engaged for the long haul.

L7: Monetization

In Layer 7, Canva's monetization strategy shines through its well-balanced free-to-paid conversion approach. The company offers robust free plans to lure users into their ecosystem, especially targeting a broad consumer base. Once users are engaged, Canva then segments them based on their willingness to pay and adds nuanced value for specific segments, such as SMB marketing teams.

They use strategically-placed paywalls on premium features to encourage a transition from individual use to team-based collaboration. However, the company takes an unobtrusive approach to paywalls, letting users organically discover them. They've found that people who encounter paywalls naturally have a higher conversion and lower churn rate, echoing successful consumer apps like Spotify. Anshul Patel, Product Growth Lead at Canva, underscores the importance of this timing, cautioning against rushing to monetize at the expense of habit formation and long-term customer value.

L8: Flywheels

In Layer 8, Canva employs a powerful growth loop strategy that goes beyond mere word of mouth and team invites. At its core, Canva has built a content marketplace where supply drives demand, creating a self-sustaining flywheel effect. Canvas's masterful use of editorially-generated, SEO-optimized (EGSO) content became an engine for user education and search visibility. Drawing a parallel to Etsy, Canva's platform encourages sellers, or in this case, professional designers, to market their templates and designs. This grassroots promotion not only enhances the growth of individual shops but also catalyzes the overall marketplace growth. The concept holds that creating such a growth loop results in a sustainable engine for exponential business expansion.

Canva's concerted effort in geographic expansion through localization and strategic partnerships further amplifies its global footprint.

Conclusion: The Iceberg in Full View

From its narrow focus on an underserved niche to its intricate multi-layered PLG strategy, Canva's growth resembles an iceberg, powerful yet mostly hidden beneath the surface. Each layer, from L1 to L8, adds another dimension to their unstoppable growth engine. It's a comprehensive playbook for any company looking to leverage a product-led growth strategy successfully. Having successfully charted the depths of the PLG iceberg, Canva serves as a compelling example of the extraordinary outcomes possible when all layers are in sync to deliver unparalleled customer value. Special thanks to Jaryd Hermann for his invaluable PLG framework and insightful analysis of Canva's growth strategy. For more wisdom on scaling businesses, be sure to follow his work at the 'How They Grow' newsletter: https://www.howtheygrow.co/.

2.4 -What Are Moats and Why Do They Matter?

The business world is like a battleground, and to win, companies need competitive advantages—or "moats"—that set them apart. In the fast-paced world of generative AI, these moats are even more crucial. Hamilton Helmer's "7 Powers: The Foundations of Business Strategy" outlines seven types of moats. Let's see how these apply to generative AI companies.

Economies of Scale

Size matters in the business world, but it's not just about being big; it's about leveraging that size for lower operational costs and improved performance. Companies like OpenAI benefit enormously from economies of scale. Each time they upgrade a model like GPT-4, the cost per user transaction drops, while the algorithm's performance improves. This establishes a virtuous cycle where growth fuels efficiency, which in turn fuels further growth.

Network Effects

Network effects create a powerful cycle: the more users a product attracts, the better the product becomes, drawing even more users. Midjourney is a standout example. It not only created a devoted user community but also opened channels for valuable user feedback. The company fortified this moat by integrating its service with Discord, expanding its reach and making it more versatile and useful with each update.

Counter-Positioning

Standing apart from the crowd can yield impressive results. Perplexity AI has eschewed the common ad-supported model of search engines in favor of a subscription-based approach. This bold move sidesteps the clutter of ads, delivering a cleaner and more focused user experience. The unique positioning makes it hard for traditional search engines to copy the approach without jeopardizing their existing business models.

Switching Costs

Imagine the hassle of moving all your belongings from one place to another; you'd think twice before doing it. Consider an AI tool designed for healthcare professionals: Its deep integration with healthcare systems and tailored patient care algorithms make it indispensable. For medical staff, switching from this tool means facing significant costs and complexities, due to its seamless connection with electronic health records. The higher the cost to switch, the more secure your customer base is.

Branding
A brand is more than a logo; it's a promise to your customers. Adobe Firefly benefits from its Adobe lineage, coming with a reputation for quality and innovation even before users have their first interaction with it. This already sets high expectations and trust in the market, making it a compelling choice for those looking to explore generative AI for creative purposes.

Cornered Resource
Sometimes, having something nobody else has can be your trump card. Harvey AI leverages its exclusive training with specialized datasets from elite law firms. These unique datasets not only improve the service but make it irreplaceable in its niche market. Financial backing from key investors adds another layer of exclusivity, further solidifying its unique market position.

Process Power
Efficient processes may not be glamorous, but they can be a secret weapon. LangChain has developed a streamlined framework that makes it easier for developers to create Natural Language Processing applications. This reduces the time from concept to market, giving LangChain an edge. Their comprehensive guides and an active developer community further enhance this efficiency, turning what is often considered a 'boring' operational aspect into a competitive strength.

In summary, building these competitive moats is not just a tactical exercise. It's a strategic imperative that requires foresight and meticulous execution. In the fast-evolving landscape of generative AI, these moats could very well determine who leads and who follows in the race for market dominance.

2.41 -Can Generative AI Companies Have Moats?

The surge in generative AI technologies, exemplified by large language models like GPT-4 and foundation models like DALL-E 3, has captivated the tech world, enabling innovative applications that quickly find product-market fit and amass users. However, this excitement is tempered by ongoing discussions among industry insiders questioning the longevity of these companies' competitive edges. The central debate revolves around whether generative AI can create defensible moats, or if the technology is destined to become commoditized, thus erasing any first-mover advantages. In this chapter, we'll navigate this contentious landscape, weighing both sides of the argument, offering our insights, and evaluating the critical role of perseverance in building enduring generative AI companies.

Red Team Perspective: Generative AI Lacks Defensible Moats

The Red Team argues that the open-source nature and abundant resources in the generative AI space make it impossible for companies to form sustainable competitive moats. Key points include:

1. **Open research, not proprietary intellectual property (IP)**: Unlike past tech innovations built on proprietary assets, generative AI thrives on openly published research, such as the transformer architecture and foundational models like GPT-4. This open access levels the playing field, evidenced by the open-source community quickly enhancing leaked models like LLaMA.

2. **Data and models are commoditized**: With the rise of large datasets like Common Crawl and easy access to computing power through cloud services, data and models are no longer scarce. Frameworks like Hugging Face Transformers further simplify the model-building process for startups.

3. **Incumbents control distribution and workflow:** Tech giants have an unbeatable edge in distribution. They can instantly deliver new AI features

to their existing installed bases while owning existing workflows, making it tough for startups to compete without the reach.

4. **Low switching costs for users**: Generative AI products usually lack personalized data storage, making it easy for consumers to switch between similar services without friction. This lack of lock-in makes it difficult for startups to establish durable user networks.

5. **Incumbents are moving quickly**: Unlike the shift to Cloud which requires a full-stack rebuild, leveraging generative AI technologies is a quick plug-and-play deployment without rearchitecting the stack. Major players are investing heavily in their own generative AI initiatives, often outpacing startups in access to proprietary data, funding, and distribution.

6. **Incumbents have the advantage of subsidizing AI by integrating it into existing revenue streams**. Companies such as Johnson and Honeywell have subsidized IoT services and integrated them into their existing business models, where they charge for devices. While startups bear the initial costs of experimentation and customer discovery to determine product-market fit, incumbents often step in and capture the significant value created by these early efforts[176].

These factors collectively make it challenging to establish defensible moats.

Blue Team Perspective: Moats Are Necessary and Achievable

The Blue Team contends that sustainable competitive moats are both attainable and crucial for generative AI companies. They argue that:

1. **Blend public and proprietary tech for unique leverage**: For example, Cohere adds proprietary clustering technology on top of GLAM (Global Language Model), and Anthropic layers in Constitutional AI alongside Claude. This hybrid model not only adds unique features but also improves performance and accuracy. It forms a technical moat that makes it difficult for open-source-only competitors to catch up.

2. **Industry-specific proprietary data are a rare asset**: Startups in the legal or healthcare sectors can leverage these unique, proprietary datasets to solve problems that generic AI models can't, creating a protective data moat.

3. **Own your vertical to form a service-based moat**: By owning a specific industry vertical and its associated end-to-end workflows, while laying proprietary vertical data on top, companies can craft AI solutions that drive significant value tailored to users' needs. This level of personalization makes customers loyal and constructs a service-based moat that's hard to penetrate. Workflow + model vertical play is the new winning formula.

4. **The power of focus and speed**: Unlike incumbents that must balance generative AI solutions' impact on multiple existing revenue streams, startups can dedicate themselves to perfecting solutions for their niche users. This singular focus lets them innovate and adapt rapidly, countering the resource advantage of bigger players.

5. **Personalization and superior UI/UX contribute to building a durable network-effect moat**: As generative AI collects more user data and feedback, it'll become smarter and more personalized. Over time, companies can engage users deeply with hyper-personalized solutions, making it costly for users to switch and forming an enduring brand loyalty. Companies, especially early movers with exceptional UI/UX, benefit as they can leverage these factors to create a lasting network-effect moat.

6. **Form a multi-layered defense through diverse data**: Systems of Intelligence™, a concept championed by Jerry Chen at Greylock, magnify the power of AI by merging diverse datasets from various systems of record. This strategy not only boosts content and metrics but also provides a safety net against vulnerabilities that could plague single-source data systems. Serving as more than an advanced AI application, this approach reshapes the very infrastructure of enterprise products, offering startups a robust, multi-faceted defense against bigger players. Salesforce's Einstein serves as a case study for this model, which has gained even more relevance with the rise of large language models[177].

The Blue Team argues that these factors combined provide a solid foundation for building sustainable and defensible competitive moats in the world of generative AI.

What Is Our View?

In a nutshell, our view is that building competitive moats in generative AI is challenging but possible. Both startups and incumbents have paths to success through focus and adaptation. However, moats need continuous upkeep due to evolving conditions. Long-term perseverance is key for companies to truly capitalize on the transformative potential of generative AI.

We believe there are credible points on both sides of this debate about competitive moats in generative AI. In the near term, the challenges the "red team" highlights are very real. Core technological insights are widely accessible. Many applications are nascent. Business models remain fluid. Incumbents are investing aggressively.

Under these conditions, building defensible moats is extremely difficult for startups. The technology alone likely won't suffice. Product-market fit can be fleeting. Distribution disadvantages are stark.

However, we concur with the "blue team" that history suggests moats do eventually emerge around new technologies. Though innovation may start out decentralized, it tends to consolidate around platforms, standards, and leading solutions over time.

We expect similar platform dynamics will arise with generative AI, but it requires persistence through the tumultuous, uncertain infancy of the technology. For those who persevere, moats should strengthen.

History suggests moats emerge around any valuable technology. Generative AI will be no exception - its capabilities are simply too transformative. Both incumbents and startups can build moats by focusing on strengths and adapting to industry shifts. For tech giants, the moat may be proprietary compute infrastructure, reach, and technical talent. Startups must play to advantages like industry expertise, customer intimacy, and product focus.

At the same time, complacency is dangerous. Changing ecosystem conditions can erode old moats while enabling new ones. Startups like Anthropic display scrappiness to leapfrog limitations[178].

So while moats exist, they must be vigilantly maintained and rebuilt as the landscape evolves. Incumbents must keep innovating despite scale. Startups need vision beyond initial use cases. Both should identify paradigm shifts - like generative AI - before point-in-time advantages fade.

This underscores that realizing the full potential of generative AI to transform industries may span years or decades, not just quarters. Its emergence resembles past platforms like electricity, microchips, software, and the internet. The companies that will lead this revolution likely have yet to be started. Many pivotal insights have yet to be discovered on the journey from narrow AI to artificial general intelligence.

Innovators who are ready for a challenging journey and have the patience and determination to learn and improve, even when they face early difficulties, will be in the best position to benefit from what is set to be one of the most groundbreaking and valuable technologies. The lasting generative AI powerhouses will emerge over time through this crucible of competition and discovery.

For now, perseverance is the name of the game.

Why Perseverance Matters?

The tech landscape is littered with innovators who fell short because they gave up too soon. Pioneering is not enough; enduring success demands the grit to navigate the winding roads of foundational technologies like microchips, PCs, and the internet. Just as today's tech giants rose to prominence not as the originators but as the most adaptable, the giants of generative AI will be those who stick around[179,180, 181].

Remember the AI winters? Once-bustling labs fell silent as DARPA funding waned in the 1970s and 1980s due to overhyped promises and underwhelming outcomes. Some stayed the course, working in near anonymity until the thaw came—laying the groundwork for today's AI spring.

So, as we stand on the cusp of the generative AI era, heed this: victory won't go to the swiftest or the most ingenious, but to those who persevere through uncertainties and setbacks. Because when the dust settles, it's the resilient that stand tall, much like Charles Darwin, who navigated uncertainties to rewrite our understanding of life. Patience and fortitude are just as essential as innovation in this long-haul journey.

Call to Action

In these wild early days of generative AI, hype understandably abounds. But realization of its full potential will be a marathon, not a sprint[182].

Lasting success for startups will require long-term thinking, not short-term speculation. We must take the time to build meaningful, ethical applications and sustainable business models, not just chase ephemeral advantages[183].

Optimism about the technology's eventual impact need not diminish cautions about the challenges in getting there. By acknowledging current limitations, we empower the best and brightest to tackle them.

So we issue a call to action. Let us embrace the journey ahead with perseverance, pragmatism and care. The foundations set today will ripple for generations to come.

Part III: Navigating the Product Career in the AI Era

In a world increasingly steered by AI, every professional domain is experiencing the seismic shifts brought on by this transformative technology. The field of product management is no exception. Product managers (PMs) have traditionally been the driving force that navigates the delicate balance between user needs, business objectives, and technological possibilities. They have often been seen as the linchpin behind a product, equipped with deep market insights and the vision needed to guide product development towards success. As AI technologies advance and become more prevalent, the landscape of product management is evolving. What will it mean to be a PM in this brave new world of AI and automation? The objective of this chapter is to help PMs understand, adapt, and thrive amidst this technological evolution. It is designed to provide both a compass and a roadmap, empowering product managers to navigate the changing terrain with confidence and skill.

3.1 -How Will Product Managers Evolve in the AI Era?

3.11 - What Does a Product Manager Do?

Product Managers are the connective tissue that hold the entire product team together. They work at the intersection of business, technology, and user experience, often acting as the 'voice of the customer' within their organization. The key responsibilities and competencies of a PM include:

1. **Setting Product Vision and Strategy**: PMs define the direction and the strategic vision for the product, ensuring alignment with business objectives and market opportunities. This requires an understanding of the competitive landscape, technological trends, and most importantly, the customer needs.
2. **Understanding Customer Needs**: PMs use various tools and techniques, like user interviews, surveys, and usability studies, to gain a deep understanding of customer needs and pain points.

3. **Defining and Prioritizing Features**: Based on the understanding of customer needs, PMs define the product features and prioritize them using frameworks like RICE (Reach, Impact, Confidence, and Effort) or MoSCoW (Must-have, Should-have, Could-have, Won't-have) to shape the product roadmap.

4. **Working with Cross-functional Teams**: PMs collaborate with engineering, design, marketing, and sales to take the product from idea to launch. They ensure seamless communication among these teams, facilitating decision-making and conflict resolution.

5. **Tracking and Improving Performance**: PMs use key performance indicators (KPIs) to track the product's success. They continuously gather and analyze user feedback to iterate and improve the product.

3.12 - Will AI Take Over Product Management Jobs?

Given that product management roles demand strong business savvy, team collaboration, and decision-making skills, does this make them resistant to being automated by AI?

A Dystopian Vision

The year is 2040. I'm Sam, one of the last remaining human product managers at Utopia, the tech conglomerate that now controls 90% of the world's commerce and communications. I just got called into a meeting with my robotic boss, the Automated Intelligence Matrix (AIM).

"Sam," AIM drones in its cold, robotic voice, "I've analyzed the data and determined that your services are no longer required. Our AI-powered Product Management Intelligence System (PMIS) has made your role obsolete."

I plead for my job, trying to explain the human subtleties of product strategy that no algorithm could ever grasp. AIM is unmoved. "I cannot justify keeping inefficient biological units like yourself on staff simply for, as you say, 'the human touch.' PMIS can analyze real-time customer data, predict usage trends, ideate solutions, and generate endless personalized product variations all without human intervention."

I protest that ideation requires creativity - something AIs still lack. AIM refutes this: "PMIS has an advanced generative ideation module. Just yesterday it came up with a novel concept for an electric jetpack rental subscription service targeted at children under 5."

My mind reels at the absurdity of jetpack-wielding toddlers zooming through the air. But AIM is already preparing my termination paperwork, beeping calmly about transitioning my duties to the PMIS hive mind.

Dragged from the office by Booglesoftazon's robo-guards, I look back and see a drone affixing my former job title on AIM's gleaming chassis: Chief Product Management Officer. AI has taken my job, and soon no human will be left doing product work.

This dystopian future may seem far-fetched today in 2023. But as AI rapidly advances, could product managers someday face obsolescence? While parts of the job like data analysis and documentation could be automated, uniquely human strengths like strategic thinking, vision, and emotional intelligence are harder to replicate. Product managers who embrace AI as an augmenting tool rather than a replacing force will maintain their relevance. But for now, the threat looms of AIM and its drone army, set to assimilate every last human PM into the singularity...

A More Realistic Perspective

The rise of AI is causing anxiety about its potential impact on human employment. According to a recent analysis by OpenAI, automation enabled by AI could affect up to 80% of jobs in the US workforce. Specifically, high-income professional and technical roles are most susceptible, with 19% of workers in those fields seeing over 50% of their tasks automated[184].

These findings align with a study by Goldman Sachs estimating that AI could substitute up to one-fourth of current occupations, which translates to around 300 million full-time jobs globally. White collar office and administrative jobs, as well as positions in law, finance, medicine, and other professions are at greatest risk[185].

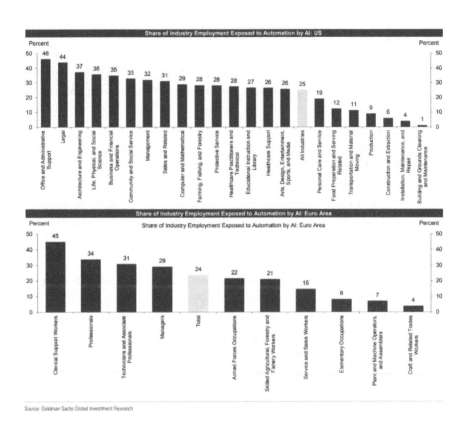

Source: Goldman Sachs Global Investment Research

While these projections may seem dire, history suggests AI automation will occur more slowly than anticipated. Corporate planning typically takes around 6 -12 months. Organizations tend to be risk averse with emerging technologies. In the short term, AI is more likely to augment human capabilities than wholesale replace jobs. For example, a banker could use AI for data analysis and rote tasks, while still relying on human skills for strategic planning with clients.

Over time, however, wide-scale displacement seems inevitable. As the capabilities of AI expand, entire occupations previously thought safe may fade away. Just as lamplighters and switchboard operators were rendered obsolete by electricity and the telephone, so, too, may today's roles like financial analysts, reporters, telemarketers, and paralegals gradually disappear thanks to increasingly intelligent machines.

New occupations also spring up to offset jobs lost to technological change. After the Industrial Revolution, while many manufacturing roles declined, entirely new fields like electrical engineering, automotive design, and aerospace emerged. AI automation may likewise give rise to new human occupations that are unimaginable today.

Preparing the workforce of the future will be critical in easing the transition period of AI adoption. Some companies are already providing training programs to help employees develop the digital skills needed to work alongside smart machines. But much larger government and private investments in education, job retraining, and social support will likely be required to reduce labor displacement pain.

Instead of a utopian future where humans are completely free from work, or a dystopian one where AI replaces us, it's more likely that people will still work similar hours as they do today, 20 years from now. Except with the help of AI, the nature of that work has shifted toward more rewarding creative and interpersonal activities. The goal then is not to resist automation altogether, but rather to shape it responsibly - so that AI empowers human potential rather than displaces it.

3.13 - What Skills Are Needed for Product Managers to Thrive in the Age of AI?

In the rapidly evolving landscape of AI, product managers must cultivate both soft and technical skills to stay ahead of the curve. From leveraging generative AI to enhance communication and strategic thinking, to mastering technical proficiencies like data analytics and prompt engineering, PMs need a multi-faceted skill set to excel.

Enhancing Soft Skills with Generative AI

Developing soft skills, or 'people skills,' is critical for product managers, especially in an AI-driven landscape. Here's how generative AI can amplify some of the key soft skills:

- **Strategic Thinking:** Generative AI aids PMs by offering instant, data-driven insights. It can analyze large volumes of customer data, market trends, forecast customer behavior, and assess competitor tactics. Despite these benefits, AI can't replace the irreplaceable value of customer discovery and a finely-tuned product sense. PMs must continually sharpen these skills to unearth deep, non-obvious user insights.

- **Creativity**: While technology progresses at a rapid pace, the core principle of crafting a compelling product remains constant. Take gaming, for example. Despite advanced graphics chips and rendering engines, what makes a game successful is the creativity behind the game's design. Technological innovations cannot substitute the critical need for creative problem-solving and innovative design. For product managers, the true challenge lies in harnessing these tools with creative, out-of-box thinking.

- **Curiosity:** Generative AI can spark a product manager's curiosity by providing easy access to expansive knowledge databases, often without requiring specialized technical know-how. With AI filtering and presenting relevant data, PMs are encouraged to ask deeper questions and explore unconventional angles, nurturing a more inquisitive mindset.

- **Communication & Storytelling:** Effective communication and storytelling are pivotal soft skills for product managers. Tools powered by generative AI, like GPT-4, can enhance these skills by offering suggestions for clearer and more engaging messaging. Building on the wisdom from Carmine Gallo's "Talk Like TED," product managers can leverage AI technologies to master the art of storytelling, thereby creating impactful narratives that resonate both their teams and stakeholders.

- **Influencing**: AI enhances a PM's persuasive prowess by providing data-driven insights to form persuasive arguments. Drawing from Chris Voss's "Never Split the Difference," which underscores the value of grasping the other party's viewpoint for effective negotiation, generative AI can sift through extensive data to yield such insights. This helps PMs understand their stakeholders more deeply and wield influence more effectively. Additionally, PMs can use AI for role-playing exercises, simulating different stakeholder perspectives to better prepare for negotiations.

- **Collaboration with Cross-Functional Stakeholders:** Generative AI streamlines teamwork by automating mundane tasks, freeing up time for more meaningful, strategic interactions. Workplace productivity tools like Microsoft Teams can further reduce the prep work needed for meetings, allowing for a more concentrated discussion. AI can handle additional tasks such as summarizing key takeaways or managing action items. Moreover, AI can also facilitate cross-cultural communication and collaboration, by breaking down language barriers with real-time translation and flagging cultural nuances, enabling clearer communication and more effective collaboration across diverse teams.

- **Ethics:** PMs must not only be aware of ethical considerations such as algorithmic bias, data privacy, and AI safety, but also take the lead in these conversations. They are accountable for ensuring their products adhere to evolving governance standards, making their role in ethical oversight increasingly critical.

In the AI era, product managers must amplify their soft skills through generative AI, all while integrating ethical considerations into their leadership and product philosophy.

Developing Technical Skills and AI Literacy

Dr. Marily Nika, AI Product Lead at Meta, a TED AI speaker, and founder of the popular AI Product Academy, believes that soon, every product manager will be an AI product manager. The landscape is quickly evolving to where AI literacy will become indispensable for product managers. Below are top 10 technical skills to develop:

1. **Data Analytics**: Understanding how to interpret data will be essential for evaluating the effectiveness of AI models.
2. **Machine Learning:** Familiarize yourself with key algorithms and understand when and how they should be applied. As chatbots and voice interfaces become more common, a basic understanding of Natural Language Processing (NLP) is invaluable.
3. **Python Programming**: Many AI tools and libraries are Python-based. Basic coding skills will help you communicate better with engineers. Leverage tools like ChatGPT Code Interpreter to speed up your learning.

4. **Data Labeling & Annotation**: Understanding how data is labeled for machine learning can offer you valuable context for how models are trained.

5. **Model Evaluation Metrics:** Learn how to gauge the effectiveness of a machine learning model through metrics like accuracy, precision, and recall.

6. **API Interactions**: Learn the basics of how APIs work to better understand how AI services can be integrated into existing products.

7. **Cloud Computing:** Understanding the cloud infrastructure will be key, as most AI tools are cloud-based.

8. **Responsible AI:** With AI, data privacy and security become even more critical. A basic understanding of cybersecurity can go a long way. Familiarize yourself with how content moderation works to ensure your products adhere to the safety and equity standards.

9. **Understanding Generative Models:** Knowing how LLMs generate text can help you set realistic expectations and identify use-cases within your product. Take time to understand how generative models can be applied to images, audio, or even code, can open new avenues for innovation.

10. **Prompt Engineering**: Mastering the art of prompt design can help you extract more targeted and useful responses from generative models. This involves understanding how to phrase prompts, set context, and apply techniques to refine output.

In the evolving world of generative AI, Amit Fulay, VP of Product for Microsoft Teams and GroupMe, emphasizes that technical literacy is becoming increasingly essential for product managers. While deep AI expertise isn't necessary, a basic understanding to facilitate effective collaboration with AI teams is crucial, mirroring the current dynamics with data science teams. The real limitation in product development is the breadth of one's imagination in conceptualizing what can be achieved. Product managers need to continually push the boundaries of innovation, challenging their AI teams with new visions for their products. It's this progressive mindset that unlocks the full potential of AI in product management, fostering cross-disciplinary synergy and driving transformative innovation.

The synergistic relationship between AI and PMs opens doors to refine and acquire essential skills, shaping the future of product management. In a speculative day-in-the-life, we explore how one PM seamlessly integrates AI to enrich her role, revealing a blueprint for the future of the profession.

Case in Point: A Speculative Day in the Life of a Product Manager in the AI Era

9 AM: Emma kicks off her morning with coffee and her AI-enhanced dashboard, which presents key metrics on customer behavior and market trends. While appreciative of the hours saved, she still dives into recent customer interviews. Her human intuition uncovers nuances the AI hasn't grasped—yet.

12 PM: By noon, Emma is in a brainstorming session with her team. She accesses an expansive knowledge database to explore potential product features and improvements. The AI tool filters out irrelevant information and suggests articles, research papers, and similar case studies from various industries that spark new ideas. The ease of this process encourages her and her team to ask deeper questions and think beyond conventional boundaries.

1 PM: Emma huddles with designers and engineers for a rapid prototyping session. Utilizing a text-prompt-to-design tool, she inputs feature descriptions. Design mocks and code snippets are generated on the fly. The engineers make some tweaks and set up the call to an API. Within minutes, they have functionable prototypes, blending her strategic vision with real-time AI outputs.

2 PM: In the afternoon, Emma drafts her product strategy narrative using generative AI, infusing it with "Talk Like TED" storytelling techniques. She also runs an AI simulation to role-play stakeholder reactions, further fine-tuning her pitch. The AI helps her preempt objections, boosting her natural ability and confidence to persuade.

4 PM: Excited to share her draft narrative, Emma lets an AI bot handle meeting invites and agendas. In the meeting, AI logs key feedback and actions, freeing the team for deep strategy talks. AI also offers real-time translations and flag cultural nuances in feedback giving and receiving, simplifying global communication.

5 PM: Before wrapping up her day, Emma reviews an ethics report generated by her Responsible AI tool. She assesses issues like data privacy violations and algorithmic bias. Emma knows she's accountable for the ethical ramifications of her products, and she takes a few moments to ensure everything is in line with industry governance standards.

As Emma shuts down her computer, she reflects on how AI has not just made her job easier, but also richer and more complex. She feels invigorated by the endless possibilities ahead and is grateful for her multifaceted skill set that enables her to harness the power of AI responsibly and effectively.

3.2 - How Can Product Managers Work Well with AI?

As AI continues to disrupt traditional roles, product managers will inevitably start to question their place in this new landscape. Reframe your perspective: see AI not as a threat but as a valuable ally, woven seamlessly into your daily activities.

Envision AI as an Enabler: the modern product manager sees AI as a liberator, a tool that automates the mundane and repetitive, allowing them to focus on the essence of their role—strategic decisions, ethical considerations, and long-term objectives. The AI doesn't usurp the human touch; it reinforces it, ensuring that product managers can invest their intellect where it truly matters.

Consider AI as your Co-Pilot: just as a seasoned aviator relies on their instruments to navigate the skies, a product manager can leverage AI as a collaborative force. It's a partnership where AI provides data-driven insights, and the manager pilots the decision-making process. This synergy allows for mutual growth, much like how different players on a team bring unique skills to the table, while keeping the human in the driver's seat.

Embrace AI as an Amplifier: AI technologies are uniquely positioned to democratize specialized knowledge and skills traditionally outside the purview of a product manager, such as coding or design. The AI amplifies human capability, rather than replacing it.

Embracing a future where AI is a cornerstone requires cultivating an inquisitive spirit, a dedication to continuous learning, and adaptability. It involves nurturing a culture of cooperation and collective wisdom that surpasses individual contributions. Most importantly, it's about looking ahead with a sense of possibility, where AI not only refines the present but also catapults professionals into a future ripe with untapped potential and boundless opportunities.

For product managers, the journey with AI is a shared voyage: AI serves as a trusted companion, propelling them not just towards achieving greater productivity but also towards a profound reimagining of their professional identity in a future defined by AI's transformative influence.

3.21 - How May Generative AI Enhance Product Development?

Generative AI Product Development Tips
(Part 1 of 2)

Phase	Generative AI Use Cases	Sample Prompts
Ideation	☐ Generate ideas based on market trends ☐ Create variations of existing products ☐ Suggest improvements to solve pain points	*"Generate a list of product ideas for urban cyclists."* *"Create variations of smartwatches focused on the elderly."* *"Suggest product improvements for remote learning tools."*
Research and Analysis	☐ Analyze market data to identify patterns ☐ Generate user interview guides ☐ Synthesize user feedback from various sources into actionable insights ☐ Develop customer journey maps ☐ Generate competitive analysis reports	*"Identify market trends in the sustainable packaging industry."* *"Summarize key points from user reviews of the latest fitness apps."* *"Create a report on the competitive landscape of meal kit services."*
Design and Prototyping	☐ Generate product design concepts ☐ Simulate user interactions with prototypes ☐ Generate user personas and scenarios for hypothesis testing ☐ Validate product ideas with multiple perspectives ☐ Seek expert advice on concepts	*"Design a concept for a modular smartphone case."* *"Simulate a user interacting with a new educational app."* *"Generate user personas and simulate a user's journey for a day using a personal finance app."* *"Assess the cultural impact and acceptance of introducing an automated drone delivery service in urban and rural environments."*
Development	☐ Draft product requirement documents ☐ Draft initial code for new features. ☐ Automate repetitive coding tasks. ☐ Offer code optimization suggestions	*"Write a script for a basic e-commerce checkout process."* *"Automate the generation of API documentation from code comments."* *"Suggest optimizations for a food delivery app's routing algorithm."*

Copyright ©2024 by Reimagined Authors Shi, Cai, and Rong

Generative AI Product Development Tips
(Part 2 of 2)

Phase	Generative AI Use Cases	Sample Prompts
Testing and Quality Assurance	☐ Generate test cases based on user stories ☐ Identify edge cases for stress testing ☐ Simulate user behavior for testing and generate synthetic data ☐ Conduct a pre-mortem to identify potential risks	*"Create test cases for a new payment processing feature."* *"Identify edge cases for a ride-sharing app during peak traffic hours."* *"Simulate customer service challenges that could arise from the introduction of a self-service AI platform in banking."*
Launch	☐ Generate marketing copy, launch checklist, social media post, product FAQs, help center articles ☐ Create targeted customer outreach emails ☐ Suggest A/B test scenarios for launch campaigns	*"Write a press release for a new gaming console."* *"Craft personalized email templates for different user segments regarding a new feature launch."* *"Generate A/B test scenarios for a digital marketing campaign for a new yoga apparel line."*
Evaluation and Iteration	☐ Analyze customer feedback for insights ☐ Generate feature improvement ideas ☐ Recommend product iteration strategies	*"Propose enhancements for a virtual event platform's networking features."* *"Develop a strategy for iterative releases for a project management tool."*
Product Visioning & Roadmap Prioritization	☐ Help with roadmap prioritization and OKR setting ☐ Explore new strategic directions, including market sizing and simulating different business and product expansion scenarios and competitive moves ☐ Compare and contrast different POVs ☐ Modeling different resource allocation scenarios to optimize team productivity and business outcomes	*"Propose OKRs for the upcoming quarter that align with our goal to increase user retention by 20%."* *"Simulate revenue impact of adding a subscription model to our existing free-to-use productivity app."* *"Simulate revenue impact of adding a subscription model to our existing free-to-use productivity app."*

Copyright ©2024 by Reimagined Authors Shi, Cai, and Rong

3.22 - How May Generative AI Accelerate PM Career Growth?

How Generative AI Can Help? Below we provide some sample prompts.

Communication

- **Influencing cross-functional stakeholders**
  - o "Help me emphasize the cross-functional perspectives, highlight the potential upsides and downsides of adopting a new project management tool."

- **Improving executive communication**
  - o "Compose an email to the executive team outlining the strategic importance of increased R&D investment. Write clearly and concisely in a way that inspires action."

- **Pitching product vision and speaking in public**
  - o "Prepare an engaging product vision pitch, in a story format and tailored for a non-technical audience, that communicates our product's future direction and its market impact."
  - o "Help me prepare for stage fright."

- **Managing uncomfortable conversations**
  - o "I need to deliver feedback to [person's name and their relationship to you] about [feedback details]. I want to make sure they feel supported yet understand my expectations. [Reasons why I feel unease about delivering this feedback]. Include 10 possible questions they might ask and how I should answer them."

- **Saying no with conviction without damaging the relationship**
  - o "Provide a firm but respectful template for rejecting a stakeholder's suggestion due to budget and scope limitations."
  - o "Let's role play this scenario. Act as if you were [person's name, their relationships to you, and their perspectives on the subject matter]."

- **Networking as an introvert**
  - o "I'm meeting [person's name], a [their role], who is interested in [their interests]. Help me prepare by providing 5 thoughtful

questions that show genuine interest in their work and help me build a connection."

- **Advocating for yourself**
  - "Examine the results my team and I have achieved. Identify 3 persuasive arguments that demonstrate our exceptional performance and value to the company. Develop a concise pitch that I can use to advocate for a promotion or pay raise during my review. Highlight how my leadership and our team's work align with the company's goals and how we've successfully tackled [specific company challenges]. The pitch should resonate with [insert decision-maker's role], addressing the key metrics and strategic objectives they care about. "

Negotiation

- **Feature scope negotiation with engineering**
  - "I'm in discussions with our engineering team about the scope of the upcoming feature set where my goal is to integrate user-requested functionalities without overextending our resources. Help me formulate 10 questions to uncover their limits on development time and complexity. Then, outline potential alternative solutions if we reach an impasse. Lastly, assist me in determining the maximum level of feature complexity we can commit to without sacrificing product quality or timeline."

- **Resource allocation negotiation for multiple product initiatives**
  - "I'm negotiating with our cross-functional leads on distributing resources among several product initiatives. My ideal outcome is to secure the necessary manpower and budget to meet our strategic goals for product X. Create a list of 10 questions that will reveal their priorities and constraints. Provide me with a set of viable alternatives in case we cannot reach a consensus. And finally, guide me to decide on the minimum resource allocation required for our product's success without jeopardizing other projects."

- **Salary and job offer negotiations**
  - "I'm preparing to negotiate my salary and job offer for a product manager position. I aim to secure a compensation package that reflects my experience and the value I can bring to the team. Craft

10 questions that will help me understand the company's salary ranges and benefits, as well as their flexibility on non-salary terms. Then, provide a list of alternative benefits or perks I could propose if the base salary offered doesn't meet my expectations. Finally, assist me in deciding on the minimum salary I should accept that would still make the offer attractive considering the entire compensation package.

Career Planning & Job Search

- **Career path exploration**
  - "Outline the typical career progression for a product manager in the tech industry, including key milestones and potential pivot points."
  - "Generate a list of roles complementary to product management that could utilize my existing skill set in a new context."

- **Resume and cover letter creation**
  - "Craft a resume that highlights my experience in product growth and my successes in cross-functional team leadership."
  - "Compose a cover letter tailored for a Senior Product Manager position that aligns my expertise with the company's mission in sustainability."

- **Job matching & skill gap analysis**
  - "Match my product management skills and experience with current job openings in the SaaS industry."
  - "Analyze my current skills against the job description for a Product Lead role at Company X and identify any gaps I need to address."

- **Interview preparation**
  - "Simulate a mock interview for a product management position, focusing on questions about managing remote teams and scaling products."
  - "Provide a list of questions I should ask prospective employers to determine the strategic importance of the product management role in their organization."

- **Personal branding**
  - "Develop a personal branding statement that encapsulates my expertise in AI-driven product development and passion for user-centric design."
  - "Suggest content ideas for my LinkedIn articles that demonstrate thought leadership in product management and UI/UX."

- **Networking for job search**
  - "Identify key individuals in my industry network to reach out to for informational interviews about emerging trends in product management."
  - "Create templates for outreach messages on professional networks that reflect my interest in product innovation and collaboration."

- **Understanding company culture fit**
  - "Generate questions I can ask during an interview to uncover insights into a company's culture and how it supports product team collaboration and innovation."

- **Offer emotional support**
  - "Provide strategies for managing stress and maintaining a positive outlook while navigating job rejections or unresponsive leads."

Collaboration & General Productivity

- **Idea generation and refinement:** AI can synthesize inputs from various team members to generate a wide range of innovative ideas, or refine existing ones, ensuring that all voices are considered and the best ideas are developed further.

- **Meeting facilitation:** AI can schedule meetings across different time zones, draft agendas based on team objectives, and create minutes and action items during meetings, which helps in keeping all team members aligned and focused on the goals.

- **Conflict resolution:** By analyzing team communications and feedback, AI can identify potential conflicts or misunderstandings early on and suggest ways to address them, helping to maintain a harmonious and productive team environment.

- **Document collaboration**: AI can help in co-authoring documents by suggesting content, editing in real-time, and managing version control, which streamlines the process of creating joint proposals, reports, or presentations.

- **Enhanced communication**: For distributed teams, AI can translate communications in real-time, allowing team members to collaborate effectively regardless of language barriers.

- **Project management**: AI can assist in assigning tasks based on team members' skills and workload, predict project timelines, and suggest adjustments to keep the team on track.

- **Automated report generation:** AI can swiftly compile and summarize daily metrics and progress reports, saving time for analysis and strategic planning.

- **Email management:** Generative AI can draft, sort, and prioritize emails, enabling the product manager to focus on high-importance communications.

- **Learning new knowledge and skills**: AI can curate and summarize relevant information, create easy-to-follow study guides and practice exercises that match your learning style, helping you pick up new skills faster.

Final Thoughts and Parting Words: Preparing for the Future of Work
As the looming presence of AI continues to infiltrate nearly every aspect of modern life, product managers stand at the forefront of a paradigm shift, where they must harness the synergy of human ingenuity and AI's analytical prowess. The key to thriving in this landscape is a robust understanding of AI trends and a willingness to evolve as a permanent beta. Product managers must use AI to proof their skill sets, blending emotional intelligence and creativity with domain knowledge to create products that resonate on a human level. By doing so, they can navigate the AI-driven job market with foresight and adaptability, ensuring their role not only remains relevant but pivotal.

The journey ahead for product managers in the AI age will be marked by continuous learning and strategic career planning. Staying attuned to AI advancements and recalibrating goals will prepare them for the inevitable industry shifts. Embracing this change, product managers can leverage AI to enhance their craft, paving the way for innovation and setting the stage for a new chapter where human and artificial intelligence collaborate to push the boundaries of what's possible in product management.

Appendix

Key Concepts in AI

- **Artificial Intelligence (AI):** AI is a discipline within computer science focused on creating systems that can reason, learn, and act autonomously, mirroring human-like thought and actions.
- **Machine Learning (ML):** ML is a subfield of AI where systems are trained on data to make predictions without being explicitly programmed.
- **Deep Learning:** Deep learning, a subset of ML, utilizes artificial neural networks inspired by the human brain to process intricate patterns.
- **Supervised Learning:** In supervised learning, models use labeled data to learn from past examples and predict future outcomes.
- **Unsupervised Learning:** Unsupervised learning techniques focus on identifying patterns in unlabeled data.
- **Reinforcement Learning:** Reinforcement learning involves agents learning from rewards or penalties resulting from their actions in an environment.
- **Reinforcement Learning from Human Feedback (RLHF):** RLHF is a machine learning approach that combines reinforcement learning techniques, such as rewards and comparisons, with human guidance to train an artificial intelligence (AI) agent.
- **Transfer learning:** Transfer learning is a technique in machine learning where knowledge gained while solving one task is reused to improve performance on a related task.
- **Fine tune:** Fine-tuning involves adapting a pre-trained model to a specific task by continuing its training on a task-specific dataset.
- **Prompting:** Prompting is the act of providing a machine learning model, especially language models, with a specific input to guide its output.
- **Generative AI:** Generative AI, rooted in deep learning, is capable of creating new content, ranging from text to images, based on its training data.
- **Generative Models vs. Discriminative Models**: Generative models produce new data instances from learned distributions. Discriminative

models classify or predict data labels. Discriminative models are a type of machine learning model, commonly used for supervised machine learning tasks and are computationally cheaper than generative models. Deep learning models can be discriminative, generative or hybrid models that combine both approaches.

Detailed Process and Methods for Assumption Validation

The process of assumption validation generally involves gathering information and data to test if our assumptions hold true in the real world. It's a learning process that helps refine our understanding of our users and the market. This process becomes more nuanced in the context of generative AI, as it involves complex technology and unique user interactions.

Desirability Validation
This involves conducting market research, user interviews, and surveys to understand the potential demand. For generative AI products, one may also consider prototyping the AI-generated content to gather user feedback on the perceived value and acceptance of such content.

Common methods to validate desirability include:
- **Market Research**: Conduct comprehensive market research to understand the demand for your product, identify target audiences, and analyze market trends. Gather data on user preferences, behaviors, and willingness to pay for such applications. In the context of generative AI products, this might involve studying trends in the AI space, consumer preferences for AI-based solutions, or the successes and failures of similar AI products in the market.
- **User Interviews**: Conduct one-on-one interviews with potential users, present the prototype, and ask open-ended questions to gather qualitative feedback. Explore users' reactions, expectations, and suggestions for improvement to validate assumptions and understand their desirability. For building generative AI products, this could involve presenting users with the AI-generated content, gauging their reactions, and soliciting feedback and suggestions for improvement.
- **Surveys and Questionnaires**: Prepare surveys or questionnaires that capture users' perceptions, preferences, and expectations of the AI-generated content. Include Likert scale questions, multiple-choice questions, and open-ended prompts to gather quantitative and qualitative data on desirability.

- **Low-fidelity Wireframes:** Create low-fidelity wireframes or mockups of the user interface to visualize the application's layout and functionality and present potential outputs from generative AI. This allows for rapid iteration and early feedback from potential users.
- **Interactive Prototypes:** Develop interactive prototypes using tools like Figma, InVision, or Adobe XD to simulate the user experience and functionality of the application. This enables users to interact with the prototype and provide feedback on the overall flow and usability.

Viability Validation

This involves market sizing, cost estimation, and financial modeling to evaluate the economic sustainability. In the case of generative AI, it's crucial to account for costs associated with data and computation.

Common methods to validate viability include:
- **Market Sizing**: This involves determining the potential size of a targeted market in terms of revenue or unit sales. For generative AI products, this might involve estimating the number of potential users or the volume of AI-generated content that could be sold or used.
- **Competitive Analysis:** Analyze the competitive landscape of existing direct and indirect competitors, assess their strengths and weaknesses, and determine the market share and potential differentiation points for the proposed application.
- **Cost Estimation**: This involves estimating the costs associated with developing, launching, and maintaining a product. In the case of generative AI products, this could include costs associated with data acquisition, computation resources, model development, and maintenance.
- **Pricing Experiments:** Conduct pricing experiments or surveys to understand users' price sensitivity and determine optimal pricing strategies. Test different pricing models, such as one-time purchases, subscriptions, or freemium models, to assess user preferences and revenue potential.
- **Financial Modeling:** Develop financial models that project revenue potential based on estimated user acquisition, retention rates, and pricing assumptions. Consider various scenarios to assess the sustainability and

profitability of the business model.In the case of generative AI products, a financial model could be used to assess the potential revenue from the sale or use of AI-generated content and weigh it against the associated costs.

Feasibility Validation

This involves prototyping and technical investigations to test if the proposed product can be built with available technology. In the realm of generative AI, this might involve developing and training preliminary AI models to understand the feasibility of generating the desired content.

Conducting a technical feasibility analysis for a generative music app for example, would involve assessing the current state of technology, data availability, technical expertise, and resources required for product development. Here's how you might go about it:

- **AI Technology Availability and Maturity**: The first step in a technical feasibility analysis is to evaluate whether the necessary AI technology for generating personalized music exists and is mature enough for commercial use. This would involve researching the state-of-the-art in music generation AI, checking academic papers, and evaluating the performance and limitations of existing models.

- **Data Availability and Access**: Generative AI models typically require large amounts of data for training. For a music generation AI, you'll need a substantial amount of quality music data that is representative of a diverse range of genres and styles in a suitable format. You must consider if you have access to such data, if it's of high quality, and if there are any legal restrictions on its use.

- **AI Model Training and Evaluation**: Training generative AI models can be resource-intensive, both in terms of computational power and time. You need to assess whether you have access to the necessary hardware or cloud infrastructure to support this. Furthermore, you will also need to establish appropriate metrics to evaluate the performance of the model and ensure it's generating musically pleasing and diverse outputs.

- **Integration with Existing Systems**: Assuming this generative music app would need to integrate with various platforms for users to distribute their AI-generated songs, you need to consider the technical requirements and limitations of these integrations. This could include APIs, data sharing agreements, and security considerations.

- **UX/UI Design** (of the low-fidelity prototype): Even though the app is an AI-based application, the user interface (UI) and user experience (UX) will play a crucial role in its acceptance by users. The UI/UX design should make it easy for users to engage with the AI, understand its outputs, and make desired adjustments to the music generated.

- **Scalability**: The system needs to be designed in a way that it can handle growing amounts of work in a graceful manner. This includes being able to accommodate more users, more music tracks, and more complex AI models in the future.

- **Maintenance and Continuous Learning**: AI models often need regular updates and improvements based on ongoing learning and user feedback. Provisions for this continuous learning and updating should be included in the technical design.

- **Data Privacy and Security**: Ensuring that user data is handled securely and in compliance with regulations (like GDPR) is a technical necessity. You need to have secure systems for data storage and handling, and these should be designed and tested as part of the technical feasibility.

By conducting a comprehensive technical feasibility analysis, the product team can identify technical challenges or constraints early in the product development process, better understand feasibility, and make informed decisions to invest in a product idea and subsequently ensure the successful implementation and deployment of the application.

Usability Validation
This involves usability testing and user experience (UX) research. For generative AI, it's not just the user interface that should be intuitive, but the AI-generated content should also be comprehensible and actionable to users.

Common methods to validate usability include:

- **Usability Testing**: This is a method where users are asked to complete tasks with the product while observers watch, listen, and take notes. The goal is to identify any usability problems, collect quantitative data on users' performance (e.g., time on task, error rates), and determine the user's satisfaction with the product. In the context of generative AI products, it's crucial to test not only the interface but also the usability of the AI-generated content. Is it understandable, useful, and applicable in the context it's presented?

- **Heuristic Evaluations**: Heuristic evaluations involve experts who evaluate the interface against established usability principles (the "heuristics"). It is a quick and easy way to find usability problems in a user interface design during the early stages of the design process. In the case of generative AI, heuristics might need to be extended to include aspects like the clarity, relevance, and utility of the AI-generated content.

Ethical Assumptions Validation

This involves risk analysis, considering ethical guidelines, and sometimes consulting with ethicists or legal experts. In the context of generative AI, one should assess the potential misuse of AI-generated content and address concerns around data privacy and consent.

Common methods to validate ethical assumptions include:

- **Ethical Risk Assessment:** This involves identifying and evaluating potential ethical risks associated with a product. These can range from misuse of data to the broader societal implications of a product. For generative AI, it's important to consider risks such as the potential for the system to generate inappropriate content, the implications of AI-created content on intellectual property rights, or the misuse of AI-generated content for deceptive purposes.

- **Privacy Impact Assessment (PIA)**: A PIA is a systematic process used to identify potential privacy risks associated with the collection, use, and disclosure of personal information. In the case of generative AI products, this would involve assessing what data the AI needs to function and how that data is being protected. Special considerations would include ensuring that the AI isn't trained on private, sensitive data without proper consent, and that any data it generates is also protected.

- **Inclusive Design Reviews**: Inclusive design seeks to ensure that products are accessible and usable by as many people as reasonably possible. For generative AI products, this would involve considering whether the AI is equally useful and usable across different user groups. In our previous example of making a generative AI app, we must consider if the app only caters to Western musical tastes, or is it equally adept at creating music from other cultures?

- **Bias and Fairness Audits**: These audits seek to uncover and address potential biases in a product's design or function. In the context of generative AI, these biases can be even more insidious, as they might be embedded in the training data and hence influence the AI-generated content. Auditing for bias would involve examining both the training data and the generated content to ensure fairness.

- **Stakeholder Consultations**: Engaging a broad range of stakeholders can provide different perspectives and uncover potential ethical issues that might have been missed. For generative AI products, this could involve consulting with user groups, legal experts, ethicists, community representatives, and others to understand different perspectives on the AI-generated content.

Acknowledgements

As the authors of "Reimagined: Building Products with Generative AI," we would like to extend our heartfelt gratitude to the individuals and communities who have been instrumental in bringing this project to fruition.

We must first acknowledge the enduring patience and encouragement of our families. Their unwavering support has been a cornerstone of our ability to undertake and complete this demanding journey.

We are immensely grateful for the foreword and guidance provided by Jia Li, whose support has been instrumental in our journey. We also thank Tamar Yehoshua, Amit Fulay, Ravi Mehta, Rob C. Wolcott, Jennifer Liu, Dr. Yangqing Jia, Laura Marino, Robert Dong, Dr. Marily Nika, Xing Yao, Tony Beltramelli, Raphael Leiteritz, Erica Van, Piyush Gupta, Phyl Terry, Kai Yang, Bill Sun, Lewis Lin, Robbie Kellman Baxter, and Pulkit Agrawal for their valuable endorsements, which have greatly amplified the reach and impact of our work.

Our gratitude also goes out to the thought leaders who contributed their valuable insights to our special Generative AI PM Learning Series: Jesse Bentert, Tony Beltramelli, Chris Lu, Barak Turovsky, Sami Ghoche, Stef Corazza, Amit Fulay, Mahesh Yadav, Polly Allen, Rupa Chartuvedi, Guillaume Boniface-Chang, Chris Butler, and Prashant Mahajan. Their participation significantly enhanced the depth and breadth of our discussions.

We are deeply grateful to our friends Elena Chen, Priya Matthew Badger, Linsey Liu, and Piyush Gupta. Priya's insightful review of our book's outline was invaluable, while Piyush generously devoted his Sunday to enhance our understanding of Large Language Models (LLMs). Their emotional support and content feedback have been a source of strength and inspiration. We are greatly appreciative of AGI House hosts Jeremy Nixon and Rocky Yu providing us the space, tools and the AI community to start our journey.

We would like to express our deep appreciation to Mike Edmonds, Dhaval Bhatt, Raymond Lee, Tony Wasserman, and Hans-Bernd Kittlaus for providing us platforms to share our content. Our guest lecture at Northwestern University and our participation in the AI Insights Summit 2023 and the International Software Product Management (ISPM) Summit 2023 offered invaluable opportunities to engage with diverse audiences, enriching both the attendees and us as authors.

We thank Matt Wagner for assisting first-time authors navigate the world of publishing. And we are thankful for the legal advice by Dawei Liu (Senior Legal Counsel at Humane) and legal consultation provided by Kai Yang, whose guidance was crucial in navigating the legal intricacies of publishing. Special thanks to our editor, Sheeva Azma, whose expertise was pivotal in editing this book under a tight timeline.

We extend our thanks to Jaryd Hermann for permitting us to use his Canva case study and the PLG iceberg framework in our book. His newsletter, *How They Grow*, is an excellent resource for growth content, and we highly recommend it to our readers for its insightful analysis and strategies. Our appreciation also extends to Dr. Marily Nika for her AI PM Bootcamp on Maven. We encourage all product managers to explore her course to bridge the knowledge from our book to real-world applications.

Lastly, we acknowledge the global community of AI enthusiasts, product managers, and technologists. Their daily innovations and challenges have been a source of continuous inspiration, making this book a reflection of our collective journey in navigating the evolving landscape of generative AI. This book is not just our creation; it is a collaborative effort made richer by the contributions of everyone mentioned and many unmentioned. Thank you all for being a part of this journey.

References

1. Oxford Reference, "Artificial Intelligence," Accessed November 26, 2023, https://www.oxfordreference.com/display/10.1093/oi/authority.201108030 95426960.
2. Josh Wardini and Ivailo Ivanov, "101 Artificial Intelligence Statistics [Updated for 2023]," TechJury, July 26, 2023, https://techjury.net/blog/ai-statistics/.
3. Pega, "What Consumers Really Think About AI," Accessed November 26, 2023, https://www1.pega.com/system/files/resources/2017-11/what-consumers-really-think-of-ai-infographic.pdf.
4. "History of Artificial Intelligence," Wikipedia, Accessed November 27, 2023, https://en.wikipedia.org/wiki/History_of_artificial_intelligence.
5. "File:Wartime picture of a Bletchley Park Bombe.jpg," Wikimedia, Accessed November 26, 2023, https://commons.wikimedia.org/wiki/File:Wartime_picture_of_a_Bletchley_Park_Bombe.jpg.
6. "Brief Academic Biography of Marvin Minsky," MIT, Accessed November 26, 2023, https://web.media.mit.edu/~minsky/minskybiog.html.
7. Thejaswin S, "AI : THE NEXT GENERATION," Medium, Accessed November 26, 2023, https://medium.com/@thejas2002/ai-the-next-generation-bc3b9e22b6e9.
8. "Shakey the Robot," Defense Advanced Research Projects Agency, Accessed November 26, 2023, https://www.darpa.mil/about-us/timeline/shakey-the-robot.
9. "Shakey," Computer History Museum, Accessed November 26, 2023, https://www.computerhistory.org/revolution/artificial-intelligence-robotics/13/289.
10. "File:Unimate pouring coffee for a woman at Biltmore Hotel, 1967.jpg," Wikimedia, Accessed November 26, 2023, https://commons.wikimedia.org/wiki/File:Unimate_pouring_coffee_for_a_woman_at_Biltmore_Hotel,_1967.jpg.
11. "https://commons.wikimedia.org/wiki/File:ELIZA_conversation.png," Wikimedia, Accessed November 26, 2023, https://commons.wikimedia.org/wiki/File:ELIZA_conversation.png.
12. "File:Deep Blue.jpg," Wikimedia, Accessed November 26, 2023, https://commons.wikimedia.org/wiki/File:Deep_Blue.jpg.
13. "File:AI for GOOD Global Summit (35173300465).jpg," Wikimedia, Accessed November 26, 2023, https://commons.wikimedia.org/wiki/File:AI_for_GOOD_Global_Summit_(35173300465).jpg.
14. Seth Baum, "2020 Survey of Artificial General Intelligence Projects for Ethics, Risk, and Policy," Global Catastrophe Risk Institute, December 31,

2020, https://gcrinstitute.org/2020-survey-of-artificial-general-intelligence-projects-for-ethics-risk-and-policy/.
15. Bergur Thormundsson, "Adoption rate for major milestone internet-of-things services and technology in 2022, in days," Statista, Jan 23, 2023, https://www.statista.com/statistics/1360613/adoption-rate-of-major-iot-tech/.
16. "Generative AI to Become a $1.3 Trillion Market by 2032, Research Finds," Bloomberg, June 01, 2023, https://www.bloomberg.com/company/press/generative-ai-to-become-a-1-3-trillion-market-by-2032-research-finds/.
17. Martin Casado and Sarah Wang, "The Economic Case for Generative AI and Foundation Models," Andreesen Horowitz, August 3, 2023, https://a16z.com/the-economic-case-for-generative-ai-and-foundation-models/.
18. Matt Bornstein, Guido Appenzeller, and Martin Casado, "Who Owns the Generative AI Platform?," Andreesen Horowitz, January 19, 2023, https://a16z.com/who-owns-the-generative-ai-platform/.
19. Kelvin Mu, LinkedIn, Accessed November 26, 2023, https://www.linkedin.com/posts/kelvinmu_artificialintelligence-generativeai-startups-activity-7043394287820017664-DLuy/.
20. Palak Goel, Jon Turow and Matt McIlwain, "The Generative AI Stack: Making the Future Happen Faster," Madrona, June 1, 2023, https://www.madrona.com/the-generative-ai-tech-stack-market-map/.
21. Palak Goel, Jon Turow and Matt McIlwain, "The Generative AI Stack: Making the Future Happen Faster," Madrona, June 1, 2023, https://www.madrona.com/the-generative-ai-tech-stack-market-map/.
22. Akash Takyar, "GENERATIVE AI: A COMPREHENSIVE TECH STACK BREAKDOWN," LeewayHertz, Accessed November 26, 2023, https://www.madrona.com/the-generative-ai-tech-stack-market-map/.
23. "The Generative AI Market Map: 335 vendors automating content, code, design, and more," CBInsights, July 12, 2023, https://www.cbinsights.com/research/generative-ai-startups-market-map/.
24. Sudowrite, Accessed November 26, 2023, https://www.sudowrite.com/.
25. Verb.ai, Accessed November 26, 2023, https://verb.ai/.
26. Copy.ai, Accessed November 26, 2023, https://www.copy.ai/.
27. Lavender.ai, Accessed November 26, 2023, https://www.lavender.ai/.
28. Viable, Accessed November 26, 2023, https://www.askviable.com/.
29. Otter.ai, Accessed November 26, 2023, https://otter.ai/.
30. Midjourney, Accessed November 26, 2023, https://midjourney.com/.
31. Stable Diffusion, Accessed November 26, 2023, https://stablediffusionweb.com/.
32. Adobe Firefly, Accessed November 26, 2023, https://www.adobe.com/sensei/generative-ai/firefly.html.
33. Lightricks, Accessed November 26, 2023, https://www.lightricks.com/.
34. Descript, Accessed November 26, 2023, https://www.descript.com/.

35. Runway, Accessed November 26, 2023, https://runwayml.com/.
36. Synthesia, Accessed November 26, 2023, https://www.synthesia.io/.
37. BHuman, Accessed November 26, 2023, https://www.bhuman.ai/.
38. Lensa, Accessed November 26, 2023, https://lensa-ai.com/.
39. Soul Machines, Accessed November 26, 2023, https://www.soulmachines.com/.
40. Kinetix, Accessed November 26, 2023, https://www.kinetix.tech/.
41. Flawless, Accessed November 26, 2023, https://www.flawlessai.com/.
42. Boomy, Accessed November 26, 2023, https://boomy.com/.
43. Riffusion, Accessed November 26, 2023, https://www.riffusion.com/.
44. Resemble.ai, Accessed November 26, 2023, https://www.resemble.ai/.
45. Metaphysic, Accessed November 26, 2023, https://www.metaphysic.ai/.
46. Perplexity.ai, Accessed November 26, 2023, https://www.perplexity.ai/.
47. You.com, Accessed November 26, 2023, https://you.com/.
48. Consensus, Accessed November 26, 2023, https://consensus.app/.
49. Twelve Labs, Accessed November 26, 2023, https://twelvelabs.io/.
50. Dropbox Team, "Introducing Dropbox Dash, AI-powered universal search, and Dropbox AI," June 21, 2023, https://blog.dropbox.com/topics/product/introducing-AI-powered-tools.
51. Glean, Accessed November 26, 2023, https://www.glean.com/.
52. Tome, Accessed November 26, 2023, https://tome.app/.
53. Beautiful.ai, Accessed November 26, 2023, https://beautiful.ai/.
54. Canva Magic Write, Accessed November 26, 2023, https://www.canva.com/magic-write/.
55. Microsoft Designer, Accessed November 26, 2023, https://designer.microsoft.com/.
56. Galileo AI, Accessed November 26, 2023, https://www.usegalileo.ai/.
57. Magician for Figma, Accessed November 26, 2023, https://magician.design/.
58. Uizard, Accessed November 26, 2023, https://uizard.io/design-assistant/.
59. Monterey AI, Accessed November 26, 2023, https://www.monterey.ai/.
60. Coldreach.ai, Accessed November 26, 2023, https://coldreach.ai/.
61. Intently.ai, Accessed November 26, 2023, https://www.getintently.ai/.
62. Regie.ai, Accessed November 26, 2023, https://www.regie.ai/.
63. Twain, Accessed November 26, 2023, https://www.twain.ai/.
64. Second Nature, Accessed November 26, 2023, https://secondnature.ai/.
65. Walnut, Accessed November 26, 2023, https://www.walnut.io/.
66. Sameday, Accessed November 26, 2023, https://www.gosameday.com/.
67. Cresta, Accessed November 26, 2023, https://cresta.com/.
68. Ada, Accessed November 26, 2023, https://www.ada.cx/.
69. ASAPP, Accessed November 26, 2023, https://www.asapp.com/.
70. Birch AI, Accessed November 26, 2023, https://birch.ai/.
71. Dialpad, Accessed November 26, 2023, https://www.dialpad.com/.
72. Forethought, Accessed November 26, 2023, https://forethought.ai/.
73. Observe.ai, Accessed November 26, 2023, https://www.observe.ai/.

74. OpenDialog, Accessed November 26, 2023, https://opendialog.ai/.
75. Harvey, Accessed November 26, 2023, https://www.harvey.ai/.
76. Spellbook, Accessed November 26, 2023, https://www.spellbook.legal/.
77. CaseText, Accessed November 26, 2023, https://casetext.com/.
78. Truewind, Accessed November 26, 2023, https://www.truewind.ai/.
79. Kick, Accessed November 26, 2023, https://kick.co/.
80. Effy AI, Accessed November 26, 2023, https://www.effy.ai/.
81. Onloop, Accessed November 26, 2023, https://www.onloop.com/.
82. Paradox, Accessed November 26, 2023, https://www.paradox.ai/.
83. Poised, Accessed November 26, 2023, https://www.poised.com/.
84. "AI Writing Assistance," Grammarly, Accessed November 26, 2023, https://www.grammarly.com/grammarlygo.
85. "Khanmigo Education AI Guide," Khan Academy, Accessed November 26, 2023, https://www.khanacademy.org/khan-labs.
86. Bernard Marr, "10 Amazing Real-World Examples Of How Companies Are Using ChatGPT In 2023," Forbes, May 30, 2023, https://www.forbes.com/sites/bernardmarr/2023/05/30/10-amazing-real-world-examples-of-how-companies-are-using-chatgpt-in-2023/.
87. https://techcrunch.com/2023/03/14/duolingo-launches-new-subscription-tier-with-access-to-ai-tutor-powered-by-gpt-4/
88. Practica. Accessed November 26, 2023, https://practicahq.com/learn.
89. Anne Lee Skates, "Five Predictions for the Future of Learning in the Age of AI," Andreesen Horowitz, https://a16z.com/2023/02/08/the-future-of-learning-education-knowledge-in-the-age-of-ai/.
90. Inflection.ai, Accessed November 26, 2023, https://inflection.ai/.
91. Replika.ai, Accessed November 26, 2023, https://replika.ai/.
92. Character.ai, Accessed November 26, 2023, https://beta.character.ai/.
93. Woebot Health, Accessed November 26, 2023, https://woebothealth.com/.
94. Wysa, Accessed November 26, 2023, https://www.wysa.com/.
95. Meeno, Accessed November 26, 2023, https://amorai.co/.
96. Lillian Weng, "LLM Powered Autonomous Agents," Lil'Log, June 23, 2023, https://lilianweng.github.io/posts/2023-06-23-agent/.
97. Kenn So, "How to create a mind," Generational, July 22, 2023, https://www.generational.pub/p/how-to-create-a-mind.
98. Kenn So, "How to create a mind," Generational, July 22, 2023, https://www.generational.pub/p/how-to-create-a-mind.
99. Jared Spataro, "Introducing Microsoft 365 Copilot – your copilot for work," Microsoft, March 16, 2023, https://blogs.microsoft.com/blog/2023/03/16/introducing-microsoft-365-copilot-your-copilot-for-work/.
100. Eirini Kalliamvakou, "Research: quantifying GitHub Copilot's impact on developer productivity and happiness," GitHub, https://github.blog/2022-09-07-research-quantifying-github-copilots-impact-on-developer-productivity-and-happiness/.

101. "Automated Transcription by Sonix.ai," Zoom App Marketplace, Accessed November 26, 2023, https://marketplace.zoom.us/apps/0-E6DKYbTfKhGwzoK3Amwg.

102. "Salesforce Announces Einstein GPT, the World's First Generative AI for CRM," Salesforce, March 6, 2023, https://www.salesforce.com/news/press-releases/2023/03/07/einstein-generative-ai/.

103. Adobe Sensei, Accessed November 26, 2023, https://www.adobe.com/sensei.html.

104. Asana, Accessed November 26, 2023, https://asana.com/sv/product/ai.

105. Adept, Accessed November 26, 2023, https://www.adept.ai/.

106. Cogram, Accessed November 26, 2023, https://www.cogram.com/.

107. Sembly, Accessed November 26, 2023, https://www.sembly.ai/.

108. The Gist, Accessed November 26, 2023, https://www.thegist.ai/.

109. Wenlong Huang et al., "VoxPoser: Composable 3D Value Maps for Robotic Manipulation with Language Models," VoxPoser, Accessed November 26, 2023, https://voxposer.github.io/voxposer.pdf.

110. 1x, Accessed November 26, 2023, https://www.1x.tech/.

111. "AI & Robotics," Tesla, Accessed November 26, 2023, https://www.tesla.com/AI.

112. Jaryd Hermann, "How Intercom Grows," Medium, November 22, 2022, https://uxplanet.org/how-intercom-grows-e15fbfea6354.

113. Barak Turovsky, "Framework for evaluating Generative AI use cases," LinkedIn, February 1, 2023, https://www.linkedin.com/pulse/framework-evaluating-generative-ai-use-cases-barak-turovsky/.

114. Joshua Xu, "0 - 1M ARR in 7 months," HeyGen, April 25, 2023, https://www.heygen.com/article/0-1m-arr-in-7-months.

115. Clément Huyghebaert, "Lessons Learned Building Products Powered by Generative AI," Buzzfeed, March 13, 2023, https://tech.buzzfeed.com/lessons-learned-building-products-powered-by-generative-ai-7f6c23bff376.

116. Noah Levin, "AI: The next chapter in design," Figma Shortcut, June 21, 2023, https://www.figma.com/blog/ai-the-next-chapter-in-design/.

117. "A guide to using prompts in Uizard," Uizard, October 19, 2023, https://uizard.io/blog/guide-to-using-prompts-in-uizard/.

118. "Word Online: 15C: Reporting Inappropriate Content," Microsoft HAX Toolkit, May 26, 2023, https://www.microsoft.com/en-us/haxtoolkit/example/word-online-g15-c-reporting-inappropriate-content/.

119. "AI Writing Assistance," Grammarly, Accessed November 26, 2023, https://www.grammarly.com/ai.

120. "Bing Search | G6: Mitigate social biases," Microsoft, Accessed November 26, 2023, https://www.microsoft.com/en-us/haxtoolkit/example/bing-search-g6-mitigate-social-biases/.

121. "Kinetix combines AI and 3D animation," VentureBeat, Accessed November 26, 2023, https://venturebeat.com/games/kinetix.

122. HAX Design Library," Microsoft HAX Toolkit, Accessed November 26, 2023, https://www.microsoft.com/en-us/haxtoolkit/library/.

123. "Google AI Principles," Google AI, Accessed November 26, 2023, https://ai.google/responsibility/principles/.

124. "Responsible AI at Grammarly," Grammarly, Accessed November 26, 2023, https://www.grammarly.com/responsible-ai.

125. "The First Principles Guiding Our Work with AI," Bill & Melinda Gates Foundation, Accessed November 26, 2023, https://www.gatesfoundation.org/ideas/articles/artificial-intelligence-ai-development-principles.

126. Lindsey Liu, "What Makes Inflection's Pi a Great Companion Chatbot," Medium, September 18, 2023, https://medium.com/@lindseyliu/what-makes-inflections-pi-a-great-companion-chatbot-8a8bd93dbc43.

127. Melanie Perkins, "Introducing Magic Studio: the power of AI, all in one place," Canva, October 4, 2023, https://www.canva.com/newsroom/news/magic-studio/.

128. "Search - Consensus," Consensus, Accessed November 26, 2023, https://consensus.app/search/.

129. Netflix Technology Blog, "Artwork Personalization at Netflix," Medium, December 7, 2017, https://netflixtechblog.com/artwork-personalization-c589f074ad76.

130. Jason Wei et al., "Chain of Thought Prompting Elicits Reasoning in Large Language Models," ArXiv:2201.11903 [Cs], October 2022, https://arxiv.org/abs/2201.11903.

131. "How to Use the OpenAI Playground with GPT-3 and GPT-4," Zapier.com, Accessed November 26, 2023, https://zapier.com/blog/openai-playground.

132. "Custom Instructions for ChatGPT," OpenAI, Accessed November 26, 2023, https://openai.com/blog/custom-instructions-for-chatgpt.

133. Dan Shipper, "Using ChatGPT Custom Instructions for Fun and Profit," Every.to, September 15, 2023, https://every.to/chain-of-thought/using-chatgpt-custom-instructions-for-fun-and-profit.

134. kolesnykbogdan, "Reddit, What Are Your Best Custom Instructions for ChatGPT?" August 1, 2023, https://www.reddit.com/r/ChatGPTPro/comments/15ffpx3/reddit_what_are_your_best_custom_instructions_for/.

135. "Https://Twitter.com/Nivi/Status/1682820984074559490," X (Formerly Twitter), July 22, 2023, https://twitter.com/nivi/status/1682820984074559490.

136. "Midjourney Prompt Helper | PromptFolder," Promptfolder, Accessed November 26, 2023, https://promptfolder.com/midjourney-prompt-helper/.

137. Elena Cavender, "Snapchat's My AI Has Users Reaching Their Snapping Point," Mashable, April 25, 2023, https://mashable.com/article/snapchat-my-ai-reactions.

138. Geoffrey Fowler, "Perspective | Snapchat Tried to Make a Safe AI. It Chats with Me about Booze and Sex," Washington Post, March 14, 2023, https://www.washingtonpost.com/technology/2023/03/14/snapchat-myai/.
139. "Differential Privacy a Privacy-Preserving System," Apple, Accessed November 23, 2023, https://www.apple.com/privacy/docs/Differential_Privacy_Overview.pdf.
140. "Concrete AI Safety Problems," OpenAI, Accessed November 23, 2023, https://openai.com/research/concrete-ai-safety-problems.
141. DeepMind, "About," Accessed November 23, 2023, https://www.deepmind.com/about.
142. "What-If Tool," GitHub, Accessed November 26, 2023, https://pair-code.github.io/what-if-tool/.
143. "Innovation and AI for Accessibility," Microsoft, Accessed November 26, 2023, https://www.microsoft.com/en-us/accessibility/innovation.
144. Partnership on AI, Accessed November 26, 2023, https://partnershiponai.org/.
145. "Democratic Inputs to AI," OpenAI, Accessed November 26, 2023, https://openai.com/blog/democratic-inputs-to-ai.
146. "Snapshot of ChatGPT model behavior guidelines," OpenAI, July 2022, https://cdn.openai.com/snapshot-of-chatgpt-model-behavior-guidelines.pdf.
147. Ethan Perez et al., "Red Teaming Language Models with Language Models," ArXiv:2202.03286 [Cs], February 2022, https://arxiv.org/abs/2202.03286.
148. "Mitigating LLM Hallucinations: A Multifaceted Approach," AI, Software, Tech, and People, Not in That Order...By X, September 16, 2023. https://amatriain.net/blog/hallucinations.
149. T. Griffin, 12 Things About Product-Market Fit, February 18, 2017, Andreessen Horowitz, https://a16z.com/2017/02/18/12-things-about-product-market-fit-2.
150. Marc Andreessen, "Pmarchive - the Only Thing That Matters," Pmarchive.com, June 25, 2007, https://pmarchive.com/guide_to_startups_part4.html.
151. Lenny Rachitsky, "How to Know If You've Got Product-Market Fit," Lenny's Newsletter, Accessed November 26, 2023, https://www.lennysnewsletter.com/p/how-to-know-if-youve-got-productmarket.
152. Rahul Vohra, "How Superhuman Built an Engine to Find Product Market Fit," FirstRound, Accessed November 26, 2023, https://review.firstround.com/how-superhuman-built-an-engine-to-find-product-market-fit.
153. "Why Use the North Star Framework?" Amplitude, Accessed November 26, 2023, https://amplitude.com/north-star/why-use-the-north-star-framework.
154. Siddharth Arora,"Most Commonly Used Metrics by Product Managers," LinkedIn, Accessed November 26, 2023,

https://www.linkedin.com/posts/siddhartharoraisb_most-commonly-used-metrics-by-product-managers-activity-7089091833028333568-TdZ_/.

155. "Kite Is Saying Farewell," Code Faster with Kite, Accessed November 16, 2022, https://www.kite.com/blog/product/kite-is-saying-farewell/.

156. Josh Howarth, "What Percentage of Startups Fail? 80+ Statistics (2022)," Exploding Topics, January 7, 2022, https://explodingtopics.com/blog/startup-failure-stats.

157. Maura Grace, "Community-Led Growth: Brand Community as a Growth Lever." NoGoodTM: Growth Marketing Agency, October 21, 2022., https://nogood.io/2022/10/21/community-led-growth/.

158. Jaryd Hermann, "How Notion Grows," Www.howtheygrow.co, Accessed November 27, 2023, https://www.howtheygrow.co/p/how-notion-grows.

159. "14 Million Users: Midjourney's Statistical Success (2023)." The Art of Dev, Accessed August 19, 2023, https://yon.fun/midjourney-statistics/.

160. "WSJ: Microsoft Is Losing Money on GitHub Copilot - Inside.com," Inside, Accessed November 27, 2023, https://inside.com/ai/posts/wsj-microsoft-is-losing-money-on-github-copilot-396183.

161. Ali Abouelatta, "Notion," Read.first1000.Co, Accessed November 27, 2023, https://read.first1000.co/p/notion?utm_source=%2Fsearch%2Fvalue%252 0curve&utm_medium=reader2.

162. Sarah Tavel, "AI Startups: Sell Work, Not Software," Www.sarahtavel.com, Accessed November 27, 2023, https://www.sarahtavel.com/p/ai-startups-sell-work-not-software.

163. "Claire vo Tackles Monetization Strategy for AI Businesses and Protecting the Super ICs in Product," Reforge, October 12, 2023, https://www.reforge.com/podcast/unsolicited-feedback/episode-7.

164. "Https://Twitter.com/DavidSacks/Status/1392274969581604866," X (Formerly Twitter), May 11, 2021, Accessed November 27, 2023, https://twitter.com/DavidSacks/status/1392274969581604866.

165. Jeff Desjardins, "5 Ways to Build a $100 Million Company," Visual Capitalist, Accessed November 26, 2023, https://www.visualcapitalist.com/5-ways-100-million-company/.

166. Hila Qu, "Five Steps to Starting Your PLG Motion," Lenny's Newsletter, Accessed November 27, 2023, https://www.lennysnewsletter.com/p/five-steps-to-starting-your-plg-motion.

167. "Growth Model Ingredients," Elena Verna, Accessed October 2023, https://www.elenaverna.com/growth-model-ingredients.

168. Leah Tharin, "What Growth Model Should We Have at Our Stage?" Www.leahtharin.com, Accessed November 26, 2023, https://www.leahtharin.com/p/what-growth-model-should-we-have.

169. "Product-Led Growth Guide Volume 2: How to Get Started with PLG," Amplitude, Accessed November 27, 2023, https://amplitude.com/resources/get-started-with-product-led-growth.

170. "Product-Led Growth Guide Volume 1: What Is PLG?" Amplitude, Accessed November 27, 2023, https://amplitude.com/resources/what-is-product-led-growth.

171. "Product Led Sales Journeys," Elena Verna, Accessed October 2023, https://www.elenaverna.com/product-led-sales-journeys.

172. "Product-Led Growth Guide Volume 2: How to Get Started with PLG," Amplitude, Accessed November 27, 2023, https://amplitude.com/resources/get-started-with-product-led-growth.

173. Lindsey Liu, "When NOT to Apply Product-Led Growth Strategy," Medium, November 8, 2021, https://medium.com/@lindseyliu/when-not-to-apply-product-led-growth-strategy-493bdecbf780.

174. Jaryd Hermann, "How Canva Grows," How They Grow, Accessed November 26, 2023, https://www.howtheygrow.co/p/how-canva-grows.

175. Jaryd Hermann, "How Canva Grows," How They Grow, Accessed November 26, 2023, https://www.howtheygrow.co/p/how-canva-grows.

176. Sangeet Paul Choudary, "How to Lose at Generative AI!" Platforms, AI, and the Economics of BigTech, October 8, 2023, https://platforms.substack.com/p/how-to-lose-at-generative-ai.

177. Jerry Chen, "The New New Moats," Greylock, June 21, 2023, https://greylock.com/greymatter/the-new-new-moats/.

178. Dylan Patel, "Google 'We Have No Moat, and Neither Does OpenAI,'" Www.semianalysis.com, May 4, 2023. https://www.semianalysis.com/p/google-we-have-no-moat-and-neither.

179. Martin Casado and Sarah Wang, "The Economic Case for Generative AI and Foundation Models," Andreesen Horowitz, August 3, 2023, https://a16z.com/the-economic-case-for-generative-ai-and-foundation-models/.

180. "Perplexity AI…," Weixin Official Accounts Platform, Accessed November 27, 2023, https://mp.weixin.qq.com/s/7HbwI-h-4NsWp388Iefuvw.

181. Tanay Jaipuria, "Startups vs Incumbents in AI," Tanay's Newsletter. March 27, 2023, https://tanay.substack.com/p/startups-vs-incumbents-in-ai.

182. GPT-4, Sonya Huang, and Pat Grady, "Generative AI's Act Two," Sequoia Capital, September 20, 2023, https://www.sequoiacap.com/article/generative-ai-act-two/.

183. TJ Nahigian and Luci Fonseca, "Base10 Blog: If You're Not First, You're Last: How AI Becomes Mission Critical," Base10.Vc, https://base10.vc/post/generative-ai-mission-critical/.

184. Tyna Eloundou et al., "GPTs Are GPTs: An Early Look at the Labor Market Impact Potential of Large Language Models," Arxiv.org, Accessed November 26, 2023, https://arxiv.org/pdf/2303.10130.pdf.

185. "One-Fourth of Current Work Tasks Could Be Automated by AI in the US and Europe," Key4Biz.it, Accessed November 26, 2023, https://www.key4biz.it/wp-content/uploads/2023/03/Global-Economics-Analyst_-The-Potentially-Large-Effects-of-Artificial-Intelligence-on-Economic-Growth-Briggs_Kodnani.pdf.

186. Torres, Teresa. 2023. "Assumption Testing: Everything You Need to Know to Get Started." Product Talk. October 18, 2023. https://www.producttalk.org/2023/10/assumption-testing/.
187. Continuous Discovery Habits. 2021. Product Talk Llc.
188. Roose, Kevin. 2023. "Silicon Valley Confronts a Grim New A.I. Metric." The New York Times, December 6, 2023, sec. Business. https://www.nytimes.com/2023/12/06/business/dealbook/silicon-valley-artificial-intelligence.html?mwgrp=a-dbar&hpgrp=c-abar&smid=url-share.
189. Allen, Hugh. n.d. "DealBook Summit 2023 Elon Musk Interview Transcript." Rev Blog. https://www.rev.com/blog/transcripts/dealbook-summit-2023-elon-musk-interview-transcript.
190. Rachitsky, Lenny. n.d. "A Guide for Finding Product-Market Fit in B2B." Www.lennysnewsletter.com. Accessed December 29, 2023. https://www.lennysnewsletter.com/p/finding-product-market-fit.
191. "AI Workflow Automation: What It Is & How to Get Started | Copy.ai." n.d. Www.copy.ai. Accessed December 31, 2023. https://www.copy.ai/blog/ai-workflow-automation.
192. "Building Generative AI Products Case Study: Copy.ai." n.d. Www.linkedin.com. Accessed December 31, 2023. https://www.linkedin.com/pulse/building-generative-ai-products-case-study-copyai-shyvee-shi-snduc/?trackingId=LlM5Up3oRFKGwZ1WFOegEQ%3D%3D.

What did you think?

Your reviews and feedback are vital in amplifying the book's impact, improving its content, and nurturing our community. We'd deeply appreciate your insights and reflections on the book. **Your voice is a key part of our growth journey!**

As a token of our gratitude for your invaluable support, you'll receive a **FREE gift package** and become a valued member of our extended book launch team!

- ✓ Checklists for Gen AI products
- ✓ Top recommended AI courses
- ✓ Community discussion guide
- ✓ PM career guide & resources
- ...and other goodies!

WRITE REVIEW

SUBMIT AN ONLINE FORM

GET A FREE GIFT

HOW TO CLAIM YOUR FREE GIFT

Use the camera app on your phone to scan the QR code to leave a review **and** visit the link below to get your free gift.

SCAN ME

AND — **VISIT https://shorturl.at/wDO18**

For business inquiries & additional ideas to enhance the book, contact us at Reimaginedauthors@gmail.com

Additional Resoucres

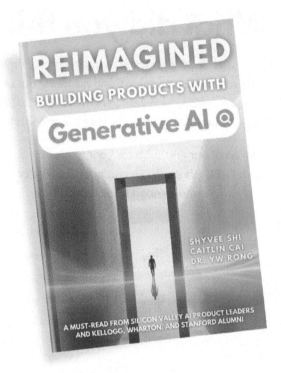

Continue your learning with us on Substack !

Product Management Reimagined

Subscribe today to hear about exclusive updates, behind-the-scenes looks into AI-powered publishing, ongoing in-depth conversations with industry experts, community gatherings, and the latest case studies in corporate innovation, product development, AI use, and career evolution!

PMReimagined.substack.com

PRODUCT MANAGEMENT LEARNING SERIES

Join 58K+ smart product builders to level up your career with industry thought leaders and practitioners!

Featuring companies:

 ... *and more!*

New LIVE sessions on Fridays and newsletters on Wednesday.

Made in the USA
Las Vegas, NV
25 January 2024

84900154R00144